Decision Engineering

Series editor

Rajkumar Roy

More information about this series at http://www.springer.com/series/5112

Annibal Parracho Sant'Anna

Probabilistic Composition of Preferences, Theory and Applications

Springer

Annibal Parracho Sant'Anna
Fluminense Federal University
Niterói, Rio de Janeiro
Brazil

ISSN 1619-5736 ISSN 2197-6589 (electronic)
ISBN 978-3-319-11276-3 ISBN 978-3-319-11277-0 (eBook)
DOI 10.1007/978-3-319-11277-0

Library of Congress Control Number: 2014949361

Springer Cham Heidelberg New York Dordrecht London

Printed on acid-free paper

Springer is part of Springer Science+Business Media (www.springer.com)

Contents

Chapter 1
Multiple Criteria Decision Analysis

Abstract A probabilistic approach to multicriteria decision problems can take in due account the uncertainty that is inevitably present in preference evaluations. Translating the preference measurements according to different criteria into probabilities of being chosen as the best alternative has two advantages. First, it makes comparable preference evaluations that come in entirely different scales. Besides, it opens probabilistic ways to automatically combine the evaluations according to the multiple criteria.

Keywords Criterion · Multicriteria decision · Preference relation

1.1 Preference Relations

The multi-criteria decision problem consists of deriving a unique preference relation between certain alternatives from evaluations of these alternatives according to a set of different criteria.

To evaluate alternatives according to a criterion means, by considering some attribute, to associate a value to each of them. According to each criterion, a different preference relation can be stated, and the problem is to derive from them a unique combination of such preference relations.

To establish a preference relation means to rank the alternatives in such a way as to be able to say if each of them is preferable to each other or not.

1.2 Classes of Decision Problems

A particular class of decision problems is that of choice. Instead of a full ranking, what is then sought is to determine among the alternatives that which is the best for a particular purpose. This is a multi-criteria decision problem if there are different criteria according to which it is possible to compare the alternatives.

A. Parracho Sant'Anna, *Probabilistic Composition of Preferences, Theory and Applications*, Decision Engineering,
DOI 10.1007/978-3-319-11277-0_1

Another case of decision problems is that of sorting, which consists of selecting from among some predetermined classes the most suitable to place an alternative. This can be seen as a choice problem in which what is chosen is, for each alternative, the most suitable from among a small number of classes. In this sorting problem, a preference between the alternatives is established whenever the predetermined classes, where such alternatives are classified, are previously ranked.

Each class is previously identified by a small set of representative alternatives. In the case of multi-criteria analysis, the vector of the evaluations of each of the class representative alternatives using the multiple criteria forms what is called a class reference profile.

1.3 Probabilities of Choice

In a probabilistic framework, attention is given to subjective aspects of the decision problem that make it impossible to evaluate the alternatives precisely. In the following chapters, an approach to take into account the presence of uncertainty in the assessments of preference and thereby to generate rules for ranking or sorting the alternatives based on probabilities of choice is presented for each decision problem.

The fact that the main interest of the decision maker—and often the sole interest—is to choose the best alternative offers a path to simplify the probabilistic modeling of the problem. In such an approach, the vectors of values of the attributes of interest give way to vectors of probabilities for presenting the best value for these attributes. Even if a ranking of all the options is desired, a better idea of the possibilities of ranks' inversion can be provided if the final ranking is derived from probabilities of being the best alternative.

Additionally, the importance of the different criteria for the choice becomes clearer if the corresponding evaluations are given in terms of probabilities of being the best according to each of them. Moreover, with all the evaluations given in the same terms, the problem of combining evaluations generated by employing different measurement standards is eliminated.

The next two chapters prepare the presentation of this probabilistic approach. After being fully developed in Chaps. 4–8, it is applied in the three last chapters in specific contexts.

1.4 Applications of the Probabilistic Approach

This probabilistic approach is applied, for instance, to the evaluation of risks, helping to detect the risks that are of higher priority. In this case, the application consists of combining risk ratings according to different sources of risk. The probabilistic composition can be applied to combine the scores of risk according to the factors of Failure Modes and Effects Analysis (FMEA): severity, frequency, and difficulty of detection of the modes of failure to generate one-dimensional risk priority probabilities.

In this context, the probabilistic approach has the advantage of allowing for flexible rules for aggregating risks generated by different factors. To increase discrimination between modes of failure related to those factors for which the scores are concentrated in a small interval, the probabilistic distributions may be modeled with a range varying with the observed range. However, to give the evaluators the option of spacing the scores only to discriminate according to factors that they judge more relevant, the probabilistic approach allows for modeling the distribution of the evaluations according to the three factors with the same range, determined by extremes previously established.

Another field of application is the assessment of productivity. The probabilistic approach can be applied in the context of evaluating the efficiency of production units employing compositions of sets of inputs to generate sets of outputs. In the probabilistic approach to this problem, the criteria can be the output/input ratios for the different pairs of input-output.

The decision may also be based on maximizing each output variable and minimizing each input variable separately. Then, a criterion will be associated with each input and each output. In this last form of modeling, the probabilities of preference according to each criterion to be computed will be, respectively, those of maximizing the revenue from the sale of each product and of minimizing the cost of the acquisition of each resource. Different treatments can then be applied to the aggregation of the two separate sets of evaluations, according to inputs and according to outputs.

The uniformization provided by the probabilistic transformation may be used to extend the possibility of application of capacities. In fact, to combine evaluations according to different criteria by the integral of Choquet with respect to a capacity in the set of criteria, the evaluations enter the computation ordered according to the values taken. This does not make sense unless the evaluations are set in a same framework. The transformation into probabilities of being the best provides this common framework.

The transformation into probabilities of being the best also has an effect of approximating to identical values the preferences for the alternatives less preferred. This feature makes the probabilistic approach useful in other contexts. For instance, it can be used to provide rough measurements to the decision attributes in applications of Rough Sets Theory with Dominance There, it allows for reducing contradictions and extracting simpler decision rules.

Chapter 2
Approaches to Criteria Combination

Abstract Two main approaches to consider the importance of the criteria in a multicriteria decision may be employed. The preferences according to each of them may be combined into a global preference by a weighted average where the importance of the criteria enters separately as weights. Otherwise the interaction between the evaluations by the multiple criteria must be taken into account when combining them and the importance of sets of criteria taken together must be computed. A simple procedure to determine the importance in the first case can be based on pairwise comparison of the criteria. A procedure to compute the importance of the sets of criteria to apply the Choquet integral in the second case may be based on pairwise comparison of preferences between distributions of probability on the space of criteria.

Keywords Weighted average · Analytic hierarchy process · Pairwise comparison · Capacity · Choquet integral

2.1 Uncertainty About Preferences

The combination of multiple criteria to decide on alternatives of action is a practical problem that every person could face at any moment. Moreover, at different times, even if the problem is formulated in the same way, the solution chosen by the same decision maker may be different. Preference depends on the criteria that are taken into account, on the importance assigned to each of them, on the relations assumed between them and, aside from all these and other features of the problem that can be objectively modeled, on subjective disturbances, which change quickly and cannot be accurately determined.

Inaccuracy affects even the simplest criteria, based on registering the presence or absence of traits considered relevant to describe each element of the set of alternatives, as in the double-entry tables of Bourdieu (1992).

A. Parracho Sant'Anna, *Probabilistic Composition of Preferences, Theory and Applications*, Decision Engineering,
DOI 10.1007/978-3-319-11277-0_2

Bourdieu suggests using double-entry tables to comparatively identify objects such as different types of educational institutions, martial arts or newspapers. One row in the table is assigned to each institution. In an inductive initial stage, a new column is opened whenever a property needed to characterize one of the individuals is found. This results in posing the issue of the presence or absence of such property for all the others. In a final stage, repetitions possibly introduced are eliminated.

By proceeding in this way, a functional or structural feature is associated with each column in such a way to retain all the features—and only those—that allow for discriminating more or less rigorously the different institutions. In these instances, preference will be related to the presence of desirable features or to the absence of undesirable ones.

In this framework, a numerical representation for the preference according to each criterion may be built by assigning the value 1 to the presence of a desirable attribute, -1 to the presence of an undesirable one and 0 to absence of the attribute, whatever its type.

When we use it for comparison purposes, this evaluation will be inaccurate because the presence or absence of even the simplest attributes may be subject to discussion. It will vary, for instance, the value that the presence or absence of a feature may have to make the object useful for the evaluator.

2.1.1 General Criteria Features

The decision becomes a bit more complex when the comparison, rather than being based on the ability to satisfy certain conditions, is based on the usefulness for a particular purpose to have different amounts of a certain attribute, where utility grows with the amount possessed. Thus, applying a criterion consists in evaluating such ability or utility.

Keeney (1992) suggests that for a candidate to effectively become a criterion an analysis of its properties should be conducted. In this examination, the criterion candidate must prove to be controllable, measurable, operational, isolable, concise, and understandable within the decision context. These and other properties must be judged with respect to the alternatives to be evaluated. These alternatives may have a high level of complexity, formed, for instance, by considering distinct results from the same experiment as more or less satisfactory.

2.1.2 Criteria Combination

Different relationships between the criteria determine different algorithms for the combination of the assessments according to them. The final results depend on the combination algorithm chosen as much as on the evaluations according to the multiple criteria.

The set of criteria must be both exhaustive, in the sense of enabling the decision maker to take into account all relevant aspects of the alternatives, and non-redundant, with each one adding some relevant aspect to discriminate between alternatives.

Roy (1996, 2005) adds to these two properties a third, called cohesiveness, as a necessary condition to have a coherent family of criteria. A family of criteria would present cohesion if a move that is not for the better in the evaluation according to any particular criteria would never lead to a move for a better general evaluation.

These and other properties are required of the set of criteria, and all have to do with the general fact that the model must represent as approximately as possible the reality, no model being able to ever completely cover all features of the decision problem. A gamma of different approaches to reduce the reality to a multi-criteria model is developed, for instance, in Greco and Ehrgott (2005) and in Ehrgott et al. (2010).

However, besides that, the form of combining the criteria depends on the goal that the decision maker has in mind. This adds to the difficulty of adequately making clear the composition rules. Modeling should not only enhance the ability to conduct the evaluation to produce an outcome that achieves the requirements of the evaluator but also the ability to explain how the values declared by the evaluator lead to the final outcome. The composition rules should allow for relating as clearly as possible the final preferences to previously exposed motives.

2.2 Weighted Averages

The classic criteria composition form, developed precisely in Keeney and Raiffa (1976), is conducted by assigning weights to the criteria and obtaining final scores as weighted averages of the measurements of preferences according to each of them. These final scores are sometimes called expected utilities.

From the point of view of making the rules clear, this form of composition has the advantage of simplicity. The model is built by defining the criteria to be taken into account and, through the weights, the importance assigned to each of them.

The idea behind this form of composition is that the decision maker starts by choosing one objective from among multiple options. Here, choosing an objective means preferring one among the multiple criteria. This choice may be not univocal but randomized, i.e., it can be given by a probability distribution of preference among the criteria.

The concept of a probability distribution will be formulated more clearly in the Appendix. This term is used here in the context of a lottery where each possible prize has a different chance of being won. In the present case, the criteria will be thought of as the prizes and the weights as their chances. This corresponds to the decision maker running a possibly biased roulette game to pick the preferred criterion.

Another way to look at this form of composition is by associating each criterion with a different evaluator and considering the averaging as a rule to satisfy the group of evaluators, with the weights corresponding to the importance assigned to each different evaluator.

A strategy to reach a distribution of weights for the criteria is to ask the decision maker to compare the criteria pairwise and, afterwards, extract from the results of these comparisons a probability for the choice of each isolated criteria.

Even establishing preferences between the elements of a pair of criteria is not free of error. However, limiting the object of each evaluation to a pair and limiting the set of outcomes of the comparison to a small set of possible results (indifference of preference for one or another, or a little more in cases where a few different degrees of preference are employed) presents considerably less difficulty than evaluating each criterion directly against some fixed pattern.

2.2.1 The Analytic Hierarchy Process

Saaty (1980) developed an elegant, though laborious, method to find the weights for the desired criteria as part of a methodology named Analytic Hierarch Process (AHP). It involves the pairwise comparison of the criteria using a scale of values for this comparison, with a criterion being at most 9 times more important than any other.

When performing this pairwise comparison, one must keep in mind that the effect of the weights depends on the different scales on which the evaluations according to the two criteria will be measured. Thus, the comparison between the weights implicitly involves a comparison between these scales. This inner scale adjustment may be avoided only if the application of all the criteria is conceived in such a way as to involve the same scale.

To tackle this problem, Saaty proposes to start the modeling by prioritizing criteria conceived in an abstract form instead of derived from the analysis of the observed attributes of the alternatives. He first defines the criteria and compares their importance. Arranging the goals, attributes, issues, and stakeholders in a hierarchy, AHP provides an overall view of the complex relationships inherent to the situation and helps the decision maker assess whether the issues in each level are of the same order of magnitude so that he can compare homogeneous elements accurately (Saaty 1990). By this approach, only when the criteria are applied to compare the alternatives by the values of its attributes do the scales on which these attributes are effectively measured appear.

A different way to address this problem of scaling is to replace each vector of attribute measurements that appear naturally with the probabilities of the different alternatives presenting the best value for such measurements. When establishing, in the next step, the priorities for the use of each of these vectors of probabilities, we have, at the same time, values measured on the same scale and conceptual criteria built on a concrete basis to compare the alternatives. The prioritization of the criteria thus defined can be made on a sounder basis than if we start from abstract concepts.

2.2.2 AHP Tools

The most noticeable feature of AHP is the form employed to address the inconsistencies arising from the pairwise comparisons. The relative preferences are registered in a square matrix M, where the ij-th entry, m_{ij}, measures the ratio between the preference for the j-th and the i-th criterion. Thus, the m_{ij} are positive numbers with

$$m_{ij} = 1/m_{ji}.$$

A square matrix with these properties is called a positive reciprocal matrix.

This matrix of preference ratios is consistent if and only if, not only

$$m_{ij} * m_{ji} = m_{ii} = 1$$

for every pair (i, j) but also, for every triple (j_1, j_2, j_3),

$$m_{j1j2} * m_{j2j3} = m_{j1j3}.$$

Obviously, given a row or a column of a reciprocal matrix, consistency determines the rest of the matrix. However, when informing the preference ratios for each pair of criteria on the scale from 1 to 9, the difficulty in evaluating abstract criteria leads to inaccuracies in such a way that it is not expected that these reciprocal matrices will be consistent in practice.

The Analytic Hierarchy Process (AHP) is designed to allow for inconsistencies due to the fact that, in making judgments, people are more likely to be cardinally inconsistent than cardinally consistent (Saaty 2003). The decision makers are not able to estimate precisely measurement values even with a tangible scale, and the difficulty is worse when they address abstract concepts.

If a reciprocal matrix is consistent, all its rows and columns are proportional to each other. This means that they span a linear space of dimension 1. In other words, the rank of consistent reciprocal matrices is equal to 1.

By the rank of a matrix we mean the dimension of the space generated by their columns (or their rows), i.e., the maximal number of linear independent columns (or rows).

In addition to the concept of rank, the concepts of the trace of a square matrix and their eigenvalues and eigenvectors play an important role in the weighting of the criteria in AHP.

The eigenvalues of a matrix M are the real or complex numbers λ such that, for some vector v,

$$Mv = \lambda v.$$

The eigenvectors are those vectors v such that, for some eigenvalue λ,

$$\mathrm{Mv} = \lambda \mathrm{v}.$$

The trace of a matrix is the sum of its eigenvalues. It is also equal to the sum of the diagonal elements. For reciprocal matrices, since all diagonal elements are equal to 1, the trace is equal to the number of rows or columns. Because the rank of any consistent reciprocal matrix is 1, its non-null eigenvectors are all in the same direction and, consequently, the non-null eigenvalue is that number of rows or columns. In particular, it is a real number.

In contrast, if the matrix is inconsistent, it has negative eigenvalues; thus, its highest eigenvalue is larger than the number of rows and columns. A detailed proof of this result may be found in Saaty (1990).

If the matrix of pairwise preferences between criteria is consistent, the vector of weights is the normalized eigenvector of the matrix. Saaty proposes then, to deal with inconsistency, to take as the vector of weights the unitary eigenvector associated with the highest eigenvalue, employing the value of this highest eigenvalue to decide if the level of inconsistency in the matrix is sufficiently small.

For high levels of inconsistency, the pairwise comparison of the criteria must be revised.

Saaty employs a measure of consistency called the Consistency Index, which is based on the deviation of the highest eigenvalue to m, the number of criteria:

$$\mathrm{CI} = (\lambda\mathrm{max} - \mathrm{m})/(\mathrm{m} - 1).$$

This index may also be seen as the negative average of the other eigenvalues of the inconsistent matrix.

After knowing the Consistency Index, the next question is how to use it. Saaty proposed to use this index by comparing it with an appropriate threshold. To determine such an appropriate threshold, one can employ the Random Consistency Index, RI, an index obtained by examining reciprocal matrices randomly generated by Vargas (1982) using the scale 1/9, 1/8 … 1, … 8, 9. The random consistency index, computed as the average of a sample of 500 matrices, for a number of criteria varying from 3 to 10, is shown in Table 2.1.

Then, Saaty employs what is called the Consistency Ratio, which is a comparison between the observed Consistency Index and the Random Consistency Index, or, formally,

$$\mathrm{CR} = \mathrm{CI}/\mathrm{RI}.$$

Table 2.1 Random consistency index

n	3	4	5	6	7	8	9	10
RI	0.58	0.90	1.12	1.24	1.32	1.41	1.45	1.49

If the value of the Consistency Ratio is smaller or equal to 10 %, the inconsistency is acceptable. If the Consistency Ratio is greater than 10 %, the subjective judgment must be revised.

2.2.3 Example of AHP Application

This subsection deals with the problem of choice of a car model. Suppose the space of alternatives formed of 20 models and six criteria based on a satisfactory answer for the presence or absence of seven attributes: beauty, comfort, gas consumption, power, acquisition price, reliability and safety.

Table 2.2 presents the ratios a given decision maker consider to more adequately reflect the preference between the seven criteria corresponding to the presence or absence of each of the seven attributes.

The highest eigenvalue for this positive reciprocal matrix is 7.335. So, its consistency index is (7.335 − 7)/(7 − 1) = 0.056. The consistency ratio is 0.056/1.32 = 0.042 < 0.1. So, the inconsistency is acceptable and a unitary eigenvector corresponding to this eigenvalue will be used as the vector of weights. Given by this unitary eigenvector, the weights are those in Table 2.3.

Table 2.4 describes the evaluation of 20 models according to the seven criteria directly based on these attributes.

Table 2.2 Criteria pairwise evaluation

	Beauty	Comfort	Consumption	Power	Price	Reliability	Safety
Beauty	1	1/3	1/5	1	1/3	1/9	1/9
Comfort	3	1	1/3	3	3	1/5	1/5
Consumption	5	3	1	3	5	1/3	1/3
Power	1	1/3	1/3	1	1/3	1/9	1/9
Price	3	1/3	1/5	3	1	1/7	1/7
Reliability	9	5	3	9	7	1	1
Safety	9	5	3	9	7	1	1

Table 2.3 Criteria weights

Beauty	Comfort	Consumption	Power	Price	Reliability	Safety
0.03	0.08	0.15	0.03	0.05	0.33	0.33

Table 2.4 Cars evaluations

	Beauty	Comfort	Consumption	Power	Price	Reliability	Safety
Car1	1	1	1	0	1	1	0
Car2	1	1	1	0	1	0	1
Car3	1	1	1	0	0	1	1
Car4	1	1	0	1	0	1	1
Car5	1	1	0	1	0	1	0
Car6	1	1	0	1	0	0	1
Car7	1	1	0	0	1	1	1
Car8	1	1	0	0	1	1	0
Car9	1	1	0	0	0	1	1
Car10	1	0	1	1	0	1	1
Car11	1	0	1	0	1	0	1
Car12	1	0	0	1	1	1	1
Car13	1	0	0	1	0	1	1
Car14	1	0	0	0	1	1	1
Car15	0	1	1	1	1	1	0
Car16	0	1	1	1	0	1	1
Car17	0	1	1	0	1	1	1
Car18	0	1	0	1	0	1	1
Car19	0	0	1	1	0	1	1
Car20	0	0	0	1	1	1	1

Comparing the weighted averages generated multiplying the value 0 or 1 assigned to the model according to each criterion by the weight of the criterion and adding the products, Car 17, with a score of 0.94, would be chosen.

2.3 Capacities

The classical approach to the composition of multiple criteria, described in the preceding section, employs weighted averages of the evaluations according to the multiple criteria. This form of composition is justified if the decision can be thought of as a two-stage structure: first, one of the criteria is chosen, with the chance of being chosen depending on the importance that the decision maker wants to give it; then, the chosen criterion is applied alone. In that case, the probability of an alternative being chosen is determined by the Total Probability Theorem, provided in the Appendix, as the weighted average of its probabilities of being chosen according to each criterion with weights given by the probabilities of the criteria being chosen in the first stage.

In this case, to build the composition algorithm, possible correlations between the events corresponding to being preferred by the different criteria need only be taken into account in determining the distribution of weights among the criteria. That is, this correlation must be taken into account in the initial stage of weighting the criteria: the probability of each one being chosen must reduce the likelihood of choosing all others positively correlated with it.

However, the problem cannot always be formulated in this manner and in general the determination of weights for the weighted average is inefficient by not taking into account these correlations. A more general form of composition that draws attention to the need to consider the possible presence of correlations is to replace the weighted average of a probability distribution by the Choquet integral with respect to a capacity (Choquet 1953).

To use this new form of composition of preferences the criteria must be comparable, i.e., the preference measurement according to the various criteria must employ the same scale, or scales between which a precise relationship is known. This problem of comparability is eliminated if the preferences are given as probabilities of being the best alternative, the scale, then, being always that of the probability of being the best.

2.3.1 Choquet Integral

To make expected utility models more flexible, additive subjective probabilities are replaced by non-additive probabilities, or capacities.

Capacities may be used to model different types of behavior. Most decision makers, for example, overestimate small and underestimate large probabilities. Furthermore, most decision makers prefer decisions where more criteria are combined rather than decisions based on less available information. These behaviors cannot be expressed through an additive model.

A (normalized) capacity on the finite set of criteria S is a set function $\mu\colon 2^S \rightarrow [0, 1]$ satisfying the three properties:

(1) $\mu(\emptyset) = 0$ (a set function satisfying this property is also called a cooperative game),
(2) $\mu(S) = 1$ (normality),
(3) $\forall A, B \in 2^N$, $[A \subseteq B \Rightarrow \mu(A) \leq \mu(B)]$ (monotonicity).

Thus, a capacity is a monotonic (normalized) cooperative game.

The capacity μ on S is said to be additive if

$$\mu(A \cup B) = \mu(A) + \mu(B)$$

for all disjoint subsets A and B of S.

Capacities generalize probabilities in the sense that an additive capacity is a probability.

The Choquet integral of $x = (x_1, \ldots, x_m)$, an R^{m+} valued function, with respect to the capacity μ on $S = \{1, \ldots, m\}$ is defined as:

$$C_\mu(x) = \sum_{j=1}^{m} (x\tau(j) - x\tau(j-1))\mu(\{\tau(j), \ldots, \tau(m)\}),$$

for τ, a permutation on S such that

$$x\tau(1) \leq x\tau(2) \leq \cdot \leq x\tau(m-1) \leq x\tau(m) \text{ and } x\tau(0) = 0.$$

Let $x: S \rightarrow R^+$ and μ a capacity. The Choquet integral of x with respect to μ satisfies

$$C\mu(x) = \sum_{i=1}^{m} x(\tau(i))[\mu(A\tau(i)) - \mu(A\tau(i+1))]$$

for $A\tau(i) = \{\tau(i), \ldots, \tau(m)\}$ for every i from 1 to m, and $A\tau(m+1) = \phi$.

A fundamental property of the Choquet integral is that

$$C_\mu(1_A) = \mu(A), \forall A \subseteq S,$$

for 1_A, the indicator of A, the function x defined by

$$x(i) = 1 \text{ if } i \in A \text{ and } x(i) = 0 \text{ otherwise.}$$

The expected value of a function x with domain S with respect to a probability P in the finite space S is the weighted average

$$\sum\nolimits_{i \in S} x(i)P(i).$$

Thus, the Choquet integral with respect to a capacity extends the expected value with respect to a probability.

But this definition makes sense only if $x_{\tau(i)}$ and $x_{\tau(j)}$, for the different possible values of i and j are commensurable. Commensurability of the measures of preference according to different criteria means that they make us able to compare the results of the evaluations according to the different criteria. This property holds for the case of evaluations according to the criteria in S given in terms of probabilities of being the best.

To compute a capacity μ, the modeler needs to define the 2^n coefficients corresponding to the capacities of the 2^n subsets of S. Modeling the capacity by means of its Moebius transform may simplify this task.

For μ a capacity on S, the Moebius transform of μ is the function ν: P(S) → R defined by

$$\nu(A) = \sum_{B \subseteq A} (-1)^{|A-B|} \mu(B), \forall A \in 2^S$$

The Moebius transform determines the capacity by:

$$\mu(A) = \sum_{B \subseteq A} \nu(B).$$

The determination of the capacity may employ the Penrose-Banzhaf or Shapley interaction indices (Grabisch and Roubens 1999) for limited levels of iteration.

Given a capacity μ on S, the Penrose-Banzhaf joint index for any subset A ⊆ S is given by (Penrose 1946; Banzhaf 1965)

$$\text{Banzhaf}(A) = 2^{-\#(S \setminus A)} \sum_{\boldsymbol{K} \subseteq \boldsymbol{S} \setminus \boldsymbol{A}} \sum_{\mathbf{L} \subseteq \mathbf{A}} (-1)^{\#(A-L)} \mu(\boldsymbol{K} \cup \boldsymbol{L}),$$

for # the cardinality function, i.e., the function that associates to each set the number of elements in it.

Analogously the Shapley joint index is defined by

$$S\mu(A) = \sum_{(K \subseteq S \setminus A)} [(\#\ (S \setminus A \setminus K))!(\#\ (K))!/(\#\ (S \setminus A) + 1)!] \sum_{L \subseteq A} (-1)^{\#(A \setminus L)} \mu(K \cup L)$$

For an isolated criterion i, $S_\mu(\{i\})$ is called the Shapley value (Shapley 1953).

The capacity μ is said to be k-additive, for a positive integer k, if its Moebius transform ν satisfies (Grabisch 1997):

(1) $\forall T \in 2^S, \nu(T) = 0$ if $\#(T) > k$,
(2) $\exists B \in 2^S$ such that $\#(B) = k$ and $\nu(B) \neq 0$.

By assuming 2-additivity, the complexity of the problem of determining the capacity is reduced. The capacity can then be determined employing only the coefficients μ({i}) and μ({i, j}) for i and j ∈ S.

Necessary and sufficient conditions for 2-additivity are:

(1) $\sum_{\{i,j\} \subseteq S} \mu(\{i,j\}) - (m-2)\sum_{i \in S} \mu(\{i\}) = 1$(normality),
(2) $\mu(\{i\}) \geq 0, \forall i \in S$ (nonnegativity) and
(3) $\forall A \subseteq S$ with $\#(A) \geq 2, \forall k \in A, \sum_{i \in A \setminus \{k\}} (\mu(\{i,k\}) - \mu(\{i\})) \geq (\#(A) - 2)\mu(\{k\})$ (monotonicity).

For a 2-additive capacity μ, the Shapley value of an isolated criterion i is given by

$$S_\mu(\{i\}) = \sum_{K \subseteq S \setminus \{i\}} [(\#(S \setminus K) - 1)!(\#(K))!/\#(S)!]\mu(K \cup \{i\}) - \mu(K)$$

or

$$S_\mu(\{i\}) = \mu(\{i\}) + 1/2 \sum\nolimits_{j \in S \setminus \{i\}} I\mu_{ij}$$

for

$$I\mu_{ij} = \mu(\{i, j\}) - \mu(\{i\}) - \mu(\{j\}).$$

$I\mu_{ij}$ represents an interaction between i and j, in the sense that

$I\mu_{ij} = 0$ corresponds to independence between i and j;

$I\mu_{ij} > 0$ means some complementarity between i and j, i.e., for the decision maker, both criteria have to be satisfactory in order to get a satisfactory alternative; and

$I\mu_{ij} < 0$ means some substitutability or redundancy between i and j, i.e., for the decision maker, the satisfaction of one of the two criteria is sufficient to have a satisfactory alternative.

With this notation, for any $x = (x_1, \ldots, x_m)$, the Choquet integral of x with respect to the 2-additive capacity μ is given by:

$$C\mu(x_1, \ldots, x_m) = \sum_{i=1}^{m} S_\mu(i)x_i - 1/2 \sum_{i,j \in S} I\mu_{ij}|x_i - x_j|.$$

2.3.2 Example of Application of the Choquet Integral

An example of application of the concept of capacity and the Choquet integral may be constructed by revising the car models choice problem presented to show how to use the AHP approach.

Essentially, a car has no value if it does not move. Thus a model evaluated as unsatisfactory from the point of view of power suffers from a basic limitation. In that sense, the value of the presence of any of the other attributes depends on the presence of power. To take this into account, a capacity might be employed to improve that study. It would derive from the weights there employed, which assumed additivity, the capacity of any set of criteria that includes power, but would assign a null capacity to any set that does not include power. This would result in a final score of zero for those cars which were assigned a value of zero with respect to power.

Table 2.5 Combined Car Scores

	Capacity score	Rank	Additive score	Rank
Car1	0	16	0.64	15
Car2	0	16	0.64	15
Car3	0	16	0.92	2.5
Car4	0.8	4	0.8	7
Car5	0.47	10.5	0.47	19.5
Car6	0.47	10.5	0.47	19.5
Car7	0	16	0.82	6
Car8	0	16	0.49	18
Car9	0	16	0.77	9
Car10	0.87	2	0.87	4
Car11	0	16	0.56	17
Car12	0.77	5.5	0.77	9
Car13	0.72	8	0.72	13
Car14	0	16	0.74	11.5
Car15	0.64	9	0.64	15
Car16	0.92	1	0.92	2.5
Car17	0	16	0.94	1
Car18	0.77	5.5	0.77	9
Car19	0.84	3	0.84	5
Car20	0.74	7	0.74	11.5

After this change, Car17 would no longer be chosen on the basis of the evaluations in Table 2.4. It would be replaced by Car16. The score of Car16 would be $0 * (0.08) + 1 * (1 - 0.08) = 0.92$.

Table 2.5 presents the final scores for the composition employing this capacity.

2.3.3 Pairwise Comparisons in the Space of Sets of Criteria

The strategy of pairwise comparisons of sets of criteria may be employed to determine the capacity. This strategy becomes more feasible if it is not needed to quantify the preference, but only to tell whether there is indifference between the elements of the pair, and, otherwise, which one is preferred.

This simplification in the comparison can be applied at the cost of complicating the objects of comparison, the decision maker being called to compare not pairs of criteria, but pairs of distributions of weights on the set of criteria. Under these conditions and certain conditions of rationality that constitute the basis of the Expected Utility Theory of von Neumann and Morgenstern (1944), the relation of preference between pairs uniquely determines the distribution of weights desired.

The work of von Neumann and Morgenstern extends that of Daniel Bernoulli, in the 18th century for the utility of money, and has further extensions designed to deal with much complex sets than the finite set of criteria that is the object of analysis here.

To clarify the hypotheses of von Neumann and Morgenstern is first necessary to formulate precisely the concepts. The necessary concepts are made clear in the next section. Immediately after, is presented the simple version of the result of von Neumann and Morgenstern (1944) here employed.

2.3.4 Binary Relations

A binary relation $\succeq$ on a set C is any subset of the Cartesian product C X C, i.e., any set of ordered pairs of elements of C. To denote that an ordered pair (c_1, c_2) of elements of C belongs to the binary relation $\succeq$, we write $c_1 \succeq c_2$ and say that c_1 precedes c_2.

A binary relation $\succeq$ is complete on C if and only if, for all c_1 and c_2 of C, at least one of the ordered pairs (c_1, c_2) and (c_2, c_1) belongs to $\succeq$, that means, $c_1 \succeq c_2$ or $c_2 \succeq c_1$.

A binary relation $\succeq$ i s anti-symmetric if and only if

$c_1 \succeq c_2$ and $c_2 \succeq c_1$ only if $c_1 = c_2$.

A binary relation $\succeq$ on C is transitive if and only if,

for all c_1, c_2 and c_3 of C, if $c_1 \succeq c_2$ and $c_2 \succeq c_3$, then $c_1 \succeq c_3$.

A binary relation $\succeq$ is an order relation on C if it is anti-symmetric, transitive and complete on C.

A binary relation $\succeq$ is a preference relation on C if it is transitive and complete on C. So, order relations are preference relations, but these need not be anti-symmetric. For those criteria for which $c_1 \succeq c_2$ and $c_2 \succeq c_1$, it will be said that according to $\succeq$ the decision maker is indifferent between c_1 and c_2 and it will be used the notation $c_1 \sim \succeq c_2$.

A distribution of weights on a set C is any positive function u with domain C, i.e., any subset of the Cartesian product CXR^+ such that for every element x of C there is a unique positive number y for which $(x, y) \in u$. Instead of $(x, y) \in u$ is usually employed the notation $y = u(x)$ or is said that y is the weight of x, the preference value of x or the utility of x. To simplify the arguments, is usually assumed that u is a probability.

The von Neumann and Morgenstern theory involves extending the preference relations on C to preference relations on the set D(C) of distributions of weights on C. Let us denote by $\succeq$ the extension of $\succeq$ to D(C). For the theorem to hold, these relations must present, besides transitivity and completeness, other properties of continuity, monotonicity, substitutability and decomposability.

Continuity means that for any weight distributions **p**, **q** and **r** on C, with $\mathbf{p} \succ \mathbf{q}$ and $\mathbf{q} \succ \mathbf{r}$, there is a real $\alpha \in [0, 1]$ such that $\mathbf{q}(c) = \alpha\, \mathbf{p}(c) + (1 - \alpha)\mathbf{r}(c)$ for any $c \in C$.

Monotonicity means that if **p** and **q** are weight distributions concentrated on $\{c_1, c_2\}$ (that means, such that $\mathbf{p}(c_1) + \mathbf{p}(c_2) = \mathbf{q}(c_1) + \mathbf{q}(c_2) = 1$), for a pair of criteria (c_1, c_2), if $c_1 \succ c_2$ and $\mathbf{p}(c_1) > \mathbf{q}(c_1)$, then $\mathbf{p} \succ \mathbf{q}$.

Substitutability means that if $\mathbf{p}(c_1) = \mathbf{q}(c_2)$ and **p** and **q** assign the same value for any other criterion c, then $c_1 \sim \succ c_2$ implies $\mathbf{p} \sim \succ \mathbf{q}$.

Decomposability employs the definition of $\mathbf{p}_\omega$: for any distribution $\boldsymbol{\omega}$ on D(C), $\mathbf{p}_\omega$ denotes the distribution on C determined by

$$\mathbf{p}_\omega(c) = \sum\nolimits_{p \in D(C)} \boldsymbol{\omega}(\mathbf{p})\, \mathbf{p}(c).$$

Decomposability holds if and only if

$$\boldsymbol{\omega}_1 = \boldsymbol{\omega}_2 \text{ is equivalent to } \mathbf{p}_{\omega 1} = \mathbf{p}_{\omega 2}.$$

These properties are not as natural as they may seem to be. But to determine weights for preference criteria, what is going to be useful from Expected Utility Theory is the representation theorem asserting that for any set of criteria C and any preference relation $\succ$ on C for which there exists an extension $\succ$ to D(C) with some properties, $\succ$ identifies a unique distribution of weights u on C such that

$u(c_1) \geq u(c_2)$ if and only if $c_1 \succ c_2$

and a unique distribution of weights u on D(C) such that, for ω_1 concentrated in c_1 and ω_2 concentrated in c_2,

$\mathbf{u}(\omega_1) \geq \mathbf{u}(\omega_2)$ if and only if $c_1 \succ c_2$.

This u is defined on D(C) by

$$\mathbf{u}(\omega) = \sum_{j=1}^{m} \omega(c_j) u(c_j) \text{ for any } \omega \in D(C).$$

If there is an outcome c_0 such that

$$u(c_0) = \sum_j \omega(c_j) u(c_j),$$

this outcome c_0 may be seen as the certainty equivalent of ω, in the sense that a distribution of weights with the unique outcome c_o has the same expected utility of ω.

When the outcomes in C have numerical values, besides computing the expected utility we can compute the expected outcome

$$\sum \omega(c)c.$$

Sometimes we can compute also the utility of the expected outcome

$$u\left(\sum \omega(c)c\right).$$

A concave utility means risk aversion and a convex utility means risk proclivity, in the sense that the utility of the distribution on D(C) that gives probability 1 to the expected outcome is, respectively, greater and smaller than its expected utility. Thus concavity represents a utility-decreasing evaluation of pure risk-bearing and convexity the contrary.

To understand the concept, suppose there are two lotteries, one that pays the expected value with certainty and another that pays the different values with their different probabilities. The utility of the first lottery is larger than the utility of the second for a risk-averse evaluation. On the other hand, giving risk a positive value would lead to a convex utility. Finally, neutrality with respect to risk would make indifferent the choice between the certain outcome and the same outcome in the average, so that

$$u\left(\sum \omega(x)x\right) = \sum \omega(x)u(x).$$

2.3.5 Example of Capacity Determination

Von Newmann and Morgenstern representation theorem provides the basis for the design of complex tools to derive the capacity of each set. Accepting the above listed conditions, instead of directly assigning a value to the set, its capacity may be derived from preferences between distributions. The evaluator will find easier to compare simple distributions involving the set than choosing a numeric value for the capacity of that set. The key idea consists of asking the decision maker appropriate questions about extreme distributions involving the set, to determine if its capacity is closer to one of two extreme values than to the other.

The procedure starts by determining the capacities of unitary sets $\{c\}$. For each such set, the evaluator answers a question about preference between two distributions: one of them assigns the value 1 to $\{c\}$; the other, a free choice between the distributions assigning the value 1 to any other unitary set in C (and consequently 0 to $\{c\}$).

If the evaluator prefers the first distribution, we conclude that the decision maker assigns to $\{c\}$ a value closer to 1 than to 0, that means a value larger than ½; if the other is preferred, we conclude that the decision maker assigns to $\{c\}$ a value smaller than ½; in the case of indifference, the quest ends, with the value ½ assigned to $\{c\}$.

Suppose the answer to this first question is a preference for the distribution that assigns the value 1 to $\{c\}$. Then we proceed by asking the preference between the distribution assigning the value 1 to {c} and another assigning the value ½ to {c} and freely assigning the value ½ to other element of C. If the evaluator prefers the first of these distributions, assigns to {c} a value closer to 1 than to ½, that means a value larger than ¾. If the other distribution is preferred, then that value is smaller than ¾. Indifference means that ¾ is the value assigned and this will be the capacity of {c}.

If the Von Neumann and Morgenstern assumptions are satisfied, indifference will appear after a while. After the capacities of the unitary sets are obtained, capacities for the sets of size two will be determined.

Suppose, for instance, the capacities of $\{c_1\}$ and $\{c_2\}$ are found to be 0.3 and 0.2. Then the capacity of $\{c_1, c_2\}$ is between 0.3 and 1. To determine this capacity, the decision maker will be asked about preference between a distribution assigning the value 1 to that set and another assigning the value 0.7 to a freely chosen subset of its complement. From preference for the value 0.7 for the complement follows that the capacity of $\{c_1, c_2\}$ for the decision maker is closer to 0.3 than to 1, what means that it is between 0.3 and 0.65. From preference for the other distribution follows a capacity from 0.65 to 1. From indifference, follow the final value of 0.65.

So, if, for instance, the interval from 0.3 to 0.65 follows from the answer obtained, the next choice will be between a distribution assigning to the set a value of 0.3 and another assigning the value 0.65. If the preference is for that assigning 0.65, we get a restriction to the interval between 0.475 and 0.65. If the preference is for the other, the capacity is between 0.3 and 0.475. If the answer is indifference, the value is 0.475. And so on.

After a logical sequence of such questions, we would eventually find a capacity representing the preference of the decision maker.

References

Banzhaf, J. F. (1965). Weighted voting doesn't work: A mathematical analysis. *Rutgers Law Review, 19*(2), 317–343.

Bourdieu, P. (1992). The Practice of Reflexive Sociology(Chap. III, pp. 217–240). In in P. Bourdieu, & L. Wacquant (Eds.), *Invitation to a reflexive sociology*. Chicago: University Chicago Press.

Choquet, G. (1953). Theory of capacities. *Annales de l'Institut Fourier, 5*, 131–295.

Ehrgott, M., Figueira, J. R., & Greco, S. (2010). *Trends in multicriteria decision analysis*. Dordrecht: Springer.

Grabisch, M. (1997). k-order additive discrete fuzzy measures and their representation. *Fuzzy Sets and Systems, 92*, 167–189.

Grabisch, M., & Roubens, M. (1999). An axiomatic approach to the concept of interaction among players in cooperative games. *International Journal of Game Theory, 28*, 547–565.

Greco, J., & Ehrgott, S. (2005). *Multicriteria Decision Analysis: state of the art survey, 3-24*. Boston: Springer.

Keeney, R, L. (1992). *Value-focused thinking: A path to creative decision making*. Cambridge, MA: Harvard University Press.

Keeney, R. L., & Raiffa, H. (1976). *Decisions with multiple objectives' preferences and value tradeoffs*. New York: Wiley.

Penrose, L. (1946). The Elementary statistics of majority voting. *Journal of the Royal Statistical Society, 109*(1), 53–57.

Roy, B. (1996). *Multicriteria methodology for decision aiding*. Dordrecht: Kluwer.

Roy, B. (2005). Paradigms and challenges. In J. Greco & S. Ehrgott (Eds.), *Multicriteria decision analysis: State of the art survey* (pp. 03–24). Boston: Springer.

Saaty, T. L. (1980). *The analytic hierarchy process: Planning, priority Setting, resource allocation*. New York: McGraw-Hill. ISBN 0-07-054371-2.

Saaty, T. L. (1990). How to make a decision: The analytic hierarchy process. *European Journal of Operational Research, 48*(1), 9–26.

Saaty, T. L. (2003). Decision-making with the AHP: Why is the principal eigenvector necessary. *European Journal of Operational Research, 145*(1), 85–91.

Shapley, L. S. (1953). A value for *n*-person games. In H. W. Kuhn, & A. W. Tucker, (Eds.), *Contributions to the theory of games* (Vol. II, pp. 307–317).

Vargas, L. G. (1982). Reciprocal Matrices with Random Coefficients. *Mathematical Modelling, 3*, 69–81.

Von Neumann, J., & Morgenstern, O. (1944). *Theory of games and economic behavior*. Princeton, NJ. Princeton University Press.

Chapter 3
The Probabilistic Approach to Preferences Measurement

Abstract As the value of an attribute only signals the position of a probability distribution of the preference, the preference for an alternative according to any criterion can be given in terms of probability of it being chosen. Composition of these probabilities to obtain a global preference combining the multiple criteria can then be performed in probabilistic terms.

Keywords Preference probability · Probability of being the best · Fuzzy logic · Fuzzy sets

3.1 Probability of Preference

This chapter presents the mechanisms for the transformation of measurements of an attribute, initially evaluated using a proper scale, into measurements of preference on a general probabilistic scale.

A natural way of evaluating the preference for an alternative is by means of the probability of this alternative being chosen from among all available alternatives.

To simplify calculations, this probability of being the best among all can be approximated by the probability of being better than a particular set of competing alternatives.

If there are many attributes, the probabilities of being better than the other alternatives according to different attributes must be combined. However, before developing the composition ideas, a framework for the identification of probabilities of preference according to each attribute must be developed. This basic framework may be similar to that of Fuzzy Sets Theory.

A. Parracho Sant'Anna, *Probabilistic Composition of Preferences, Theory and Applications*, Decision Engineering,
DOI 10.1007/978-3-319-11277-0_3

3.2 Fuzzy Sets and Fuzzy Logic

A process to derive probability distributions from punctual measurements has been employed in the Theory of Fuzzy Sets since Zadeh (1965). The idea is to replace exact numbers by intervals around them, with the pertinence of each point to the interval decreasing as these points veer away from the initial measurement. Later, to combine the probabilistic preferences, operations similar to those of Fuzzy Logic (Zadeh 1978) may be employed.

The concept of a fuzzy set was created by Lofti Zadeh in Zadeh (1965). Let X be a space of points, with a generic element of X denoted by x. A fuzzy set A in X is characterized by a membership function $\mu_A(x)$ which associates to each point in X a real number in the interval [0, 1], with the values of $\mu_A(x)$ representing the degree of membership of x to A. Thus, the nearer the value of $\mu_A(y)$ to unity, the higher the degree of membership of y to A.

This definition of a fuzzy set extends the definition of a set in the ordinary sense of the term. The degrees of membership of 0 and 1 correspond to the two possibilities of truth and false pertinence to an ordinary set, called a crisp set in the Theory of Fuzzy Sets.

3.2.1 *Fuzzy Numbers*

Membership functions for fuzzy sets can be defined in any form as long as they follow the rules of the definition of a fuzzy set. The shape of the membership function used defines the fuzzy set. For instance, a triangular membership function to determine a fuzzy interval with extremes a and b around a point M will have the form

$$\mu_A(x, a, M, b) = \begin{cases} 0, & x \leq a \\ \frac{x-a}{M-a}, & a \leq x \leq M \\ \frac{b-x}{b-M}, & M \leq x \leq b \\ 0, & b \leq x \end{cases}$$

Other way to represent a fuzzy set is by α-cuts. The membership function $\mu_A : R \rightarrow [0, 1]$ of a fuzzy set A is determined by its family of α-cuts$\{A_\alpha\}_{\alpha \in [0,\ 1]}$, which are the intervals

$$A_\alpha = \left[A^L(\alpha), A^U(\alpha)\right],$$

for

$$A^L(\alpha) = inf\{z \in R : \mu_A(z) \geq \alpha\}$$

and

$$A^U(\alpha) = sup\{z \in R : \mu_A(z) \geq \alpha\}$$

3.2.2 Fuzzy Logic

A way to define a distance between two fuzzy numbers A and B is by their α-cuts:

$$d(A,B) = \sqrt{\int_0^1 (A^L(\alpha) - B^L(\alpha))^2 d\alpha + \int_0^1 (A^U(\alpha) - B^U(\alpha))^2 d\alpha}$$

Other operations involving fuzzy sets are generalizations of the crisp sets operations of complementation, intersection and union and are called standard fuzzy set operations. The complement of a fuzzy set A with membership function μ_A is the fuzzy set A^c with membership function $\mu_A C$ defined by

$$\mu_A C(x) = 1 - \mu_A(x)$$

The intersection A∩B between the fuzzy sets A and B with membership functions $\boldsymbol{\mu}_\mathbf{A}$ and $\boldsymbol{\mu}_\mathbf{B}$ has the membership function defined by

$$\boldsymbol{\mu}_{A \cap B}(x) = \min[\boldsymbol{\mu}_\mathbf{A}(x), \boldsymbol{\mu}_\mathbf{B}(x)].$$

The union AUB of the fuzzy sets A and B with membership functions $\boldsymbol{\mu}_\mathbf{A}$ and $\boldsymbol{\mu}_\mathbf{B}$ has the membership function defined by

$$\boldsymbol{\mu}_{A \cup B}(x) = \max[\boldsymbol{\mu}_\mathbf{A}(x), \boldsymbol{\mu}_\mathbf{B}(x)].$$

This corresponds to the idea that to belong to the union is sufficient to belong to one of the sets and to belong to the intersection is necessary to belong to all the sets intersected. There are other possible generalizations.

3.3 Computation of Probabilities of Preference

In the probabilistic composition, the initial point value of an attribute is seen not as a measure of definitive preference but as signaling the position of the location parameter of a probability distribution of the preference that would occur if the value of the attribute were observed under similar circumstances in a series of evaluations of the alternative over time.

Thus, the key idea of the transformation of measurements of attributes using natural scales into probabilities of preference is to translate each measurement of the basic attribute into an interval of possible satisfaction evaluations that may occur if the alternative is evaluated in successive assessments of the preference based on that attribute.

For this notion to make sense, it is necessary to take into account the imprecision that follows from the subjectivity implicit in the application of any criterion of choice. The application of the criterion is based on the measurement of the attribute and on the possibility of an error in the transposition of that measurement into an evaluation of the satisfaction of the evaluator with the presence of the attribute at the level that the measurement indicates.

3.3.1 Computation of the Probability of Being the Best

After associating the measurements of the attribute for the n alternatives with a vector of n random variables X1, . . . , Xn, the next step is to derive from this vector the probabilities of being the best. From a joint distribution of n random variables, software like R (R© Core Team 2014) or MATLAB (Mathworks 2014) immediately produces the probability of each X_a presenting a value higher than the value of any other X_b.

A few general assumptions help to determine the joint distributions of the random variables informing the evaluations according to n criteria. One assumption that considerably simplifies the modeling of joint distribution is that of independence. It is reasonable to consider as independent the errors in the measurement of the satisfaction of the criterion resulting from each of the observed values of the attribute considered.

Other assumptions have to do with shape. Continuous distributions avoid jumps in isolated points, which are usually difficult to address. In addition to independence, the limitation to continuous distributions is recommended at this point.

Among the continuous distributions, the normal distribution is employed to represent the effect of subjective measurement deviations. With easier forms to visualize shape and allowing for simpler computation, triangular and uniform distributions may replace the normal (Sant'Anna and Sant'Anna 2001; Sant'Anna 2002). For numerical evaluations on a scale of few alternatives, a lognormal distribution for the absolute deviations has also been recommended (Martino 1970).

3.3.2 Example of Probability of Being the Best

To fix ideas consider the evaluation of gas consumption by the 20 cars of the previous chapter. Nine of them were evaluated as presenting the attribute of low gas consumption and eleven not presenting it. The randomized criterion will assign to each car measure of satisfaction by gas consumption a probability distribution that will have a mode of one for the first nine cars and of zero for the other eleven. The important point is that this will be a probability distribution, not a constant.

Let us take for this distribution the triangular form. To make the translation precise, the exact value 1 will be replaced by a triangular probability distribution on [0, 1] with mode 1 and the exact value 0 by a triangular probability distribution on [0, 1] with mode 0. The probability of a car of the first group presenting the highest value in a draw of evaluations of the 20 cars by this criterion will then be the probability of [$X_a \geq X_b$ for all b $\neq$ a from 1 to 20] for X_a and X_b independent random variables. For triangular distributions on [0, 1], corresponding to the 19 values of b, eight of them with mode 1 and eleven with mode 0, this value is 0.1054.

3.4 Combination of Probabilities of Preference

To combine the probabilities of preference according to multiple criteria into a unique score of preference, different computations of probabilities of events may be employed.

The most common way is through weighted averages. Taking the probabilities of preference according to multiple criteria as conditional probabilities and assuming a distribution of preferences among the criteria, the Theorem of Total Probability gives a final probability as a linear combination of the probabilities of choice of the alternative by the criteria with weights given by the probabilities of choice of the criteria.

The formula for the final probability of preference by alternative A will then be

$$\mathrm{p(A)} = \sum_{j=1}^{m} p_j(A)\mathrm{p}(\mathrm{B_j}),$$

where p_j(A) denotes the probability of the choice of A according to the j-th criterion and $p(B_j)$ denotes the probability of the choice of the j-th among the m criteria.

To better account for the interactions between the criteria, instead of the weighted average, the Choquet integral may be employed. Determining the preferences among the criteria by a capacity μ, the final score will be given by

$$\mathrm{C_\mu(A)} = \sum_{j=1}^{m} p_{\tau(j)}(\mathrm{A}) - p_{\tau(j-1)}(\mathrm{A})\mu(\{\tau(j), \ldots, \tau(m)\})$$

for τ, a permutation of {1, … , m} such that

$$p_{\tau(j)}(\mathrm{A}) > p_{\tau(j-1)}(\mathrm{A}) \quad \text{for every } j$$

and

$$p_{\tau(0)}(\mathrm{A}) = 0.$$

Other more intuitive forms of compositions of the preferences do not depend on weighting neither the isolated criteria nor sets of them. For instance, the global score may be given by the probability of being the preferred alternative according to every criterion or by the probability of being preferred by at least one criterion.

The event of being the preferred alternative according to every criterion is the intersection of the events of being the best according to each criterion and the event of being the preferred according to at least one criterion is the complement of the union of the events of not being preferred.

Assuming independence, the final score for alternative A is then given, in the first case, by the product $\pi_j p_j(A)$ of the probabilities of choice according to isolated criteria.

In the second case, treating the union as the complement of the intersection of the complements, the final score under independence will be given by $1 - \pi_j(1 - p_j(A))$.

To check the importance of the assumption of independence, these scores may be compared to those obtained by replacing the product by the minimum, what corresponds to assuming the hypothesis of maximum dependence or, equivalently, to the fuzzy operation of intersection described in the previous section.

Since $\min_j(1 - p_j(A))$ equals $1 - \max_j p_j(A)$, in the case of employing the probability of being the best according to at least one, the score for maximal dependence will be $\max_j p_j(A)$, what also corresponds to the operation of fuzzy union of the last section.

3.5 Example of Choice of a Car

The transformation into probabilities of being the best may be employed to replace the indicator of presence and absence of the attributes for the cars with values 0 and 1 in Table 2.4 by fractionary numbers in such a way that the sum of the values in any column equals 1.

This will make those criteria contemplating rarer attributes assign higher values for the presence of that attribute. For instance, to the presence in nine models of low gas consumption, assuming the independent triangular distributions with modes zero and one, corresponds a value of 0.1054, larger than the value of 0.096 assigned to the presence of low acquisition price, what occurs in ten models.

Table 3.1 Cars probabilities

	Beauty	Comfort	Consumption	Power	Price	Reliability	Safety
Car1	0.0705	0.0756	0.1054	0.0034	0.0960	0.0585	0.0018
Car2	0.0705	0.0756	0.1054	0.0034	0.0960	0.0016	0.0621
Car3	0.0705	0.0756	0.1054	0.0034	0.0040	0.0585	0.0621
Car4	0.0705	0.0756	0.0047	0.0881	0.0040	0.0585	0.0621
Car5	0.0705	0.0756	0.0047	0.0881	0.0040	0.0585	0.0018
Car6	0.0705	0.0756	0.0047	0.0881	0.0040	0.0016	0.0621
Car7	0.0705	0.0756	0.0047	0.0034	0.0960	0.0585	0.0621
Car8	0.0705	0.0756	0.0047	0.0034	0.0960	0.0585	0.0018
Car9	0.0705	0.0756	0.0047	0.0034	0.0040	0.0585	0.0621
Car10	0.0705	0.0025	0.1054	0.0881	0.0040	0.0585	0.0621
Car11	0.0705	0.0025	0.1054	0.0034	0.0960	0.0016	0.0621
Car12	0.0705	0.0025	0.0047	0.0881	0.0960	0.0585	0.0621
Car13	0.0705	0.0025	0.0047	0.0881	0.0040	0.0585	0.0621
Car14	0.0705	0.0025	0.0047	0.0034	0.0960	0.0585	0.0621
Car15	0.0022	0.0756	0.1054	0.0881	0.0960	0.0585	0.0018
Car16	0.0022	0.0756	0.1054	0.0881	0.0040	0.0585	0.0621
Car17	0.0022	0.0756	0.1054	0.0034	0.0960	0.0585	0.0621
Car18	0.0022	0.0756	0.0047	0.0881	0.0040	0.0585	0.0621
Car19	0.0022	0.0025	0.1054	0.0881	0.0040	0.0585	0.0621
Car20	0.0022	0.0025	0.0047	0.0881	0.0960	0.0585	0.0621

Table 3.1 presents the results of the seven probabilistic transformations to the columns of Table 2.4.

The weighted average scores generated multiplying by the weight of the criterion the value assigned to each model by each criterion and adding the products of the multiplications, along with those generated combining by the Choquet integral with respect to the capacity employed in Chap. 2, are given in Table 3.2.

The scores are followed in this table by the ranks derived from them. There is a considerable agreement between the present ranks and those of Chap. 2, but a discrepancy could be noticed in the rank of Car2, which passes from 15 to 9. In the middle of the ranking, the differences between probabilistic scores are small, what generates large differences in the ranks. If the probabilistic scores were approximated to the second decimal, then the probabilistic rank of Car2 would be 11, reflecting a tie between eleven models, from the 6th to the 16th.

Table 3.2 Probabilistic scores

	Capacity score	Rank	Additive score	Rank
Car1	0.0029	17.5	0.0488	13
Car2	0.0028	19	0.0499	9
Car3	0.0034	13	0.0641	3
Car4	0.0515	4	0.0515	7
Car5	0.0316	11	0.0316	20
Car6	0.0327	10	0.0327	19
Car7	0.0034	13	0.0536	6
Car8	0.0029	17.5	0.0337	18
Car9	0.0034	13	0.0490	12
Car10	0.0582	2	0.0608	4
Car11	0.0027	20	0.0440	17
Car12	0.0499	5	0.0503	8
Car13	0.0457	9	0.0457	16
Car14	0.0033	16	0.0477	15
Car15	0.0463	8	0.0493	11
Car16	0.0620	1	0.0646	2
Car17	0.0034	15	0.0666	1
Car18	0.0495	6	0.0495	10
Car19	0.0561	3	0.0587	5
Car20	0.0478	7	0.0482	14

References

Martino, J. P. (1970). The lognormality of Delphi estimates. *Technological FForecasts, 1*(4), 355–358.

Mathworks. (2014). *MATLAB 8.0 and statistics toolbox 8.1*. Natick-MA: The MathWorks, Inc.

R© Core Team. (2014). R: A language and environment for statistical computing. *R foundation for statistical computing*, Vienna, Austria. ISBN 3-900051-07-0. Retrieved from, http://www.R-project.org/.

Sant'Anna, A. P. (2002). Data envelopment analysis of randomized ranks. *Pesquisa Operacional, 22*(2), 203–215.

Sant'Anna, A. P., & Sant'Anna, L. A. F. P. (2001). Randomization as a stage in criteria combining. In J. L. D. Ribeiro, F. S. Fogliatto, & L. B. M. Guimarães (Eds.), *Production and distribution challenges for the 21st century* (pp. 248–256). Salvador: ABEPRO.

Zadeh, L. A. (1965). Fuzzy sets. *Information and Control, 8*, 338–353.

Zadeh, L. A. (1978). Fuzzy sets as the basis for a theory of possibility. *Fuzzy Sets and Systems, 1*, 3–28.

Chapter 4
Computation of Probabilities of Preference

Abstract The starting point to the computation of preferences by multiple criteria are vectors of objective measurements of attributes of the alternatives or evaluations in a Likert scale, or even approximate rankings. These initial values are treated as location parameters of probability distributions, for instance as modes of triangular distributions or means of normal distributions. For each criterion, the probability of a realization of the distribution representing each alternative being the best in a sample can then be computed. The final score of an alternative is obtained combining its multiple probabilities of being the best.

Keywords Preference in a Likert scale · Preference distribution modeling · Joint preference distribution · Comparison to a representative sample

4.1 Preferences Quantification

A precise measure of the preference for an alternative in a problem of choice is given by the probability of it being chosen. When using only one criterion, a strategy of three stages may be followed to compute this probability.

The first consists of obtaining for each alternative an exact measurement of an attribute that provides a basis for the choice.

In a second stage, a probability distribution is determined around each of these exact measurements, taking into account that, when a person responds to external stimuli produced by objective attributes, in order to build and organize his or her level of knowledge, the interpretation that is given to any particular stimulus is subjective and varies from person to person (Roy 2005).

The diffuse evaluations of the diverse alternatives interweave in such a way that no one of them must be preferred without consideration of its varying position with respect to each of the others. In the final step, the probabilistic decision will be taken on the basis of the probability of each alternative presenting the best measurement in a sample with an entry drawn from each of these distributions.

A. Parracho Sant'Anna, *Probabilistic Composition of Preferences, Theory and Applications*, Decision Engineering,
DOI 10.1007/978-3-319-11277-0_4

This chapter is directed to the discussion of practical aspects of each of these three stages.

4.2 The Initial Quantification of Attributes

The preference for an alternative is given by the probability of its choice. However, the measurement of preferences need not be made directly in probabilistic terms. For each criterion, there is a most natural initial form of explaining the position of each alternative with respect to the others. This natural form is sometimes based on the value of an attribute measured objectively in terms of quantity, weight, cost, speed, etc. For other criteria, it will be based on evaluations of abstract concepts and expressed in ordinary language, in terms of small, moderate or large preference.

Any form of measurement may be employed to obtain initial evaluations. However, some properties of the initial data may improve the effectiveness of the probabilistic transformation. For instance, initial standardization procedures may simplify the computation and the future interpretation of the results.

Handling data measured on the same scale facilitates comparisons. Although the probabilistic transformation standardizes the scale, a general rule for the formulation of the initial assessments also facilitates uniformization. To establish a general rule, the initial evaluations may be set on a Likert (1932) scale of five linguistic positions: "very small", "small", "moderate", "large" and "very large", represented in the sequence by the numbers 1, 2, 3, 4 and 5. This scale may be amplified to a scale of nine positions with inclusion of a position close to each extreme position and two positions flanking the central position. When the attribute is measured by a continuous variable, a discretization to such a Likert scale may be realized.

Otherwise, a starting point may be ranking the alternatives, ties allowed, and reducing subsequently the number of classes by deciding to either consider some alternatives in adjacent ranks as tied or increase distances between some others. A systematic procedure to make experts in each attribute revise the distances between adjacent alternatives, possibly reducing them to zero, may be made available to improve the quality of the initial evaluations.

4.3 Modeling the Probability Distribution

To obtain probabilistic distributions, randomness is introduced by adding disturbances with null expected values to the exact initial values. This makes the initial value a central measure, a location parameter of the distribution. The determination of the other parameters is simplified by assuming independence between disturbances affecting different alternatives and equal dispersion.

As errors in measurement due to the conceptual distance between the concrete attributes observed and the subjective satisfaction extracted from them are the source of the random disturbances, it is reasonable to model the distributions as normal distributions. However, simplifications may be considered.

If the attributes that give rise to the preferences are represented by discrete variables, the approach of Fuzzy Sets Theory, of replacing exact values by membership functions, is naturally taken. A triangular distribution will then be able to model the randomness.

However, even if continuous distributions are assumed from the beginning and no discretization of the initial values is performed, triangular distributions may be assumed instead of normal distributions.

Discretization may be avoided to explore the accuracy of the measurements when the alternatives are compared by weight, volume or other physical features. Imprecision in the preference derived from such accurate measures is still due to the subjectivity of the evaluator in evaluating the benefit or cost that results from them. In that case, as the centers of distinct distributions may be closer than in the discrete case, to maintain small the distance between distributions with close centers, a normal distribution is more adequate than a triangular distribution.

Another point to take into account in the choice of the shape of the distributions is the dependence or not of the disturbance on the value observed for the attribute. The basic approach is modeling the random component as resulting from identically distributed measurement errors, as in classical statistical models. A symmetric distribution around the location parameter will then appear.

It may be more realistic, however, to make the dispersions depend on the location, to compensate for a possible excessive deviation in the initial measurement towards the preferred side. The distributions centered closer to the maximum or the minimum observed value can spread more to the side where there is a larger range of other possible values.

To obey this principle, the modeler may employ asymmetric triangular distributions with the steepest slope for the side where the extreme is closer and a milder decline for the side where the extreme is more distant. This may be done, for instance, in the case of the Likert scale of nine levels from 1 to 9 by adopting extremes of 0 and 10 for all distributions.

Alternatively, to reduce the importance of unobserved values in one of the possible extremes, different bounds may be used for the different criteria. If the vector of observed values for the j-th attribute is $(e_{1j}, \ldots, e_{nj})$, extremes for the triangular distributions will then be fixed at $e_{0j} = \min\{e_{1j}, \ldots, e_{nj}\} - 1$ and $e_{n+1j} = \max\{e_{1j}, \ldots, e_{nj}\} + 1$.

In the symmetric case, if a normal distribution is assumed, the observed values for the other alternatives may also be taken as a basis to model a dispersion parameter, which is enough to determine a normal distribution after the mean is known. To give a reasonable chance of occurrence in any distribution for all the observed values, the standard deviation of the vector $(e_{1j}, \ldots, e_{nj})$, given by $\left(\left(\sum_{a=1}^{n}\left(e_{aj} - \sum_{b=1}^{n} e_{bj}/n\right)^2\right)/n\right)^{1/2}$, or some transformation of it, may be used as a common dispersion parameter for all of them.

4.4 Transformation in Probabilities of Being the Best

The association of distributions of possible values to exact measurements makes it possible to replace, by means of a simple computation, the vector of measurements of an attribute in a set of alternatives with a vector of probabilities of each of these alternatives being the best.

After associated a probability distribution for the evaluation of each of a set of n alternatives, the measure of preference for the a-th of these alternatives is given by the probability of, in a sample formed by taking a random realization of each of the n distributions, the value obtained for the a-th distribution being larger than all the other n − 1 values.

This development assumes that an alternative is better than other if its value is higher. If the opposite occurs, i.e., the criterion associates a larger value of the attribute with a lower preference for the alternative, instead of the score being the probability of presenting the highest value in the sample, it will be the probability of presenting the lowest one.

This may be put more formally, with $e_j = (e_{1j}, \ldots, e_{nj})$ representing again the vector of measurements of the j-th attribute in the n alternatives and $p_j = (p_{1j}, \ldots, p_{nj})$ representing the vector of probabilities of preference. In the triangular model, the probabilistic preference p_{aj} is given by the probability that, in an eventual vector of observations of n random variables $(\mathrm{E}_{1j}, \ldots, \mathrm{E}_{nj})$ with independent triangular distributions with extremes e_{0j} and e_{n+1j} and with modes respectively at $e_{1j}, \ldots, e_{nj}$,

$$E_{aj} \geq E_{bj} \quad \text{for all } b \text{ from 1 to } n.$$

Assuming, without loss of generality, the extremes $e_{0j} = 0$ and $e_{n+1j} = 1$ and assuming independence between the disturbances that affect the evaluation of different alternatives, the probability of maximization, for that alternative with evaluation e_{aj} is directly given by

$$\sum_{l=0}^{n} \int_{e_{lj}}^{e_{l+1j}} \left\{ \prod_{b=1}^{l-1} \left[1 - (1-x)^2/(1-e_{bj}) \right] \prod_{b=l}^{n} \left(x^2/e_{bj} \right) \right\} \mathrm{f_a(x)dx}.$$

In these products, replace by 1 the factors with $b = a$, as well as the first product if $l = 0$ or 1 and the second if $l = n$ or $n + 1$. The integration is with respect to the triangular density $\mathrm{f_a}$ given by

$$\mathrm{f_a(x)} = 2(1-\mathrm{x})/(1-\mathrm{e_{aj}}) \quad \text{for } a < l$$

and

$$f_a(x) = 2x/e_{aj} \quad \text{for } a > l.$$

The transformation in probabilities of maximizing the preference allows for translating from the linear scale of ranks to a more realistic scale with concentration of significant preferences on a small number of alternatives. While a few most-preferred alternatives receive clearly distinct probabilistic evaluations, probabilities of preference close to zero are given to the other alternatives.

Thus, the transformation from ranks to probabilities of being the best or the worst alternative brings an additional benefit, besides those advantages inherent to taking uncertainty into account: it increases the distance between the most important alternatives. This answers to the practical need of the evaluators to distinguish the most important alternatives with higher accuracy, while distinctions between the least preferred ones may be less reliable. Barzilai et al. (1987), Tryantaphilou et al. (1994) and Brugha (2000), among others, present good reasons to prefer nonlinear scales with such form.

4.5 Comparison to a Sample

The calculation of probabilities of being the most preferred alternative involves comparison to all the competing alternatives, even to those with the worst ratings. This ensures greater resistance to the influence of errors in the evaluation of some alternatives, but, if the number of alternatives is too large, the probabilities of preference become too small. If the comparison were made to a representative sample with a small number of alternatives, the computation would be simplified.

An undesirable side effect of this sampling approach is that then the sum of the scores will no longer be 1, as will happen if they are given by the probabilities of being the best in the whole population of alternatives. In fact, as a result of comparison to a smaller sample, all the scores become higher. But a comparatively higher score still means higher preference.

The simplification may be performed without losing the desired robustness. For that, the small sample must be built in a representative way. It may be constituted, for instance, of the deciles or the quartiles of the set of evaluations observed.

Table 4.1 Sample of three reference alternatives

Quartiles	Beauty	Comfort	Consumption	Power	Price	Reliability	Safety
1st quartile	0	0	0	0	0	1	1
Median	1	1	0	1	0.5	1	1
3rd quartile	1	1	1	1	1	1	1

An advantage of comparing to a sample of quantiles is that, by fixing the kind of quantile, the scale of the evaluations is standardized. For instance, in the case of the nine deciles, forming with the alternative being evaluated a total of ten, it may be

Table 4.2 Sample of nine reference alternatives

Deciles	Beauty	Comfort	Consumption	Power	Price	Reliability	Safety
1	0	0	0	0	0	0	0
2	0	0	0	0	0	1	0.5
3	0.5	0	0	0	0	1	1
4	1	1	0	0	0	1	1
5	1	1	0	1	0.5	1	1
6	1	1	1	1	1	1	1
7	1	1	1	1	1	1	1
8	1	1	1	1	1	1	1
9	1	1	1	1	1	1	1

Table 4.3 Probabilities of maximization in a sample of 4

	Beauty	Comfort	Consumption	Powerer	Price	Reliability	Safety
Car1	0.3214	0.3214	0.4405	0.3214	0.4153	0.25	0.25
Car2	0.3214	0.3214	0.4405	0.0595	0.4153	0.25	0.0357
Car3	0.3214	0.3214	0.4405	0.0595	0.4153	0.0357	0.25
Car4	0.3214	0.3214	0.4405	0.0595	0.093	0.25	0.25
Car5	0.3214	0.3214	0.1119	0.3214	0.093	0.25	0.25
Car6	0.3214	0.3214	0.1119	0.3214	0.093	0.25	0.0357
Car7	0.3214	0.3214	0.1119	0.3214	0.093	0.0357	0.25
Car8	0.3214	0.3214	0.1119	0.0595	0.4153	0.25	0.25
Car9	0.3214	0.3214	0.1119	0.0595	0.4153	0.25	0.0357
Car10	0.3214	0.3214	0.1119	0.0595	0.093	0.25	0.25
Car11	0.3214	0.0595	0.4405	0.3214	0.093	0.25	0.25
Car12	0.3214	0.0595	0.4405	0.0595	0.4153	0.0357	0.25
Car13	0.3214	0.0595	0.1119	0.3214	0.4153	0.25	0.25
Car14	0.3214	0.0595	0.1119	0.3214	0.093	0.25	0.25
Car15	0.3214	0.0595	0.1119	0.0595	0.4153	0.25	0.25
Car16	0.0595	0.3214	0.4405	0.3214	0.4153	0.25	0.0357
Car17	0.0595	0.3214	0.4405	0.3214	0.093	0.25	0.25
Car18	0.0595	0.3214	0.4405	0.0595	0.4153	0.25	0.25
Car19	0.0595	0.3214	0.1119	0.3214	0.093	0.25	0.25
Car20	0.0595	0.0595	0.4405	0.3214	0.093	0.25	0.25

taken as a standard for comparison that, for a completely homogeneous population, the probabilities of preference will be equal to 0.1, this number becoming a pattern for comparison of easy memorization.

To explain the possibility of loss in information due to the reduction of the size of the sample, the example of 20 cars of Chap. 2 may be enlightening. With 20 alternatives there is no need to simplify the analysis, as the sample of 20 is perfectly manageable. But it is interesting to verify that if the reduction to nine deciles is applied no substantial difference occurs. On the other side, comparing only to the quartiles, the result is considerably changed.

The samples of three and nine reference units for each criterion are, respectively, the quartiles given in Table 4.1 and the deciles given in Table 4.2. The probabilities of maximization of the preference when the comparison is to the quartiles are presented in Table 4.3 and the probabilities of preference with respect to the deciles in Table 4.4.

The correlation between the vector of ranks resulting from the comparison to the reduced sample, shown in Table 4.5, and that resulting from the comparison to the full sample in Chap. 3 reaches a value of 0.98 if the reduction is for the sample of

Table 4.4 Probabilities of maximization in a sample of 9

	Beauty	Comfort	Consumption	Power	Price	Reliability	Safety
Car1	0.1383	0.1394	0.1866	0.0134	0.1843	0.1105	0.0079
Car2	0.1383	0.1394	0.1866	0.0134	0.1843	0.0064	0.1226
Car3	0.1383	0.1394	0.1866	0.0134	0.0174	0.1105	0.1226
Car4	0.1383	0.1394	0.0181	0.1599	0.0174	0.1105	0.1226
Car5	0.1383	0.1394	0.0181	0.1599	0.0174	0.1105	0.0079
Car6	0.1383	0.1394	0.0181	0.1599	0.0174	0.0064	0.1226
Car7	0.1383	0.1394	0.0181	0.0134	0.1843	0.1105	0.1226
Car8	0.1383	0.1394	0.0181	0.0134	0.1843	0.1105	0.0079
Car9	0.1383	0.1394	0.0181	0.0134	0.0174	0.1105	0.1226
Car10	0.1383	0.0102	0.1866	0.1599	0.0174	0.1105	0.1226
Car11	0.1383	0.0102	0.1866	0.0134	0.1843	0.0064	0.1226
Car12	0.1383	0.0102	0.0181	0.1599	0.1843	0.1105	0.1226
Car13	0.1383	0.0102	0.0181	0.1599	0.0174	0.1105	0.1226
Car14	0.1383	0.0102	0.0181	0.0134	0.1843	0.1105	0.1226
Car15	0.01	0.1394	0.1866	0.1599	0.1843	0.1105	0.0079
Car16	0.01	0.1394	0.1866	0.1599	0.0174	0.1105	0.1226
Car17	0.01	0.1394	0.1866	0.0134	0.1843	0.1105	0.1226
Car18	0.01	0.1394	0.0181	0.1599	0.0174	0.1105	0.1226
Car19	0.01	0.0102	0.1866	0.1599	0.0174	0.1105	0.1226
Car20	0.01	0.0102	0.0181	0.1599	0.1843	0.1105	0.1226

Table 4.5 Final scores and ranks

	Scores by deciles	Ranks by deciles	Scores by quartiles	Ranks by quartiles
Car1	0.0920	15	0.2968	1
Car2	0.0955	11	0.2183	14
Car3	0.1215	3	0.2183	14
Car4	0.1006	7	0.2729	3.5
Car5	0.0628	20	0.2314	8
Car6	0.0663	19	0.1607	19.5
Car7	0.1046	6	0.1607	19.5
Car8	0.0667	18	0.2397	7
Car9	0.0962	10	0.1690	18
Car10	0.1155	4	0.2236	10.5
Car11	0.0851	17	0.2598	5
Car12	0.0986	8	0.1973	17
Car13	0.0903	16	0.2266	9
Car14	0.0942	13	0.2105	16
Car15	0.0925	14	0.2187	12
Car16	0.1220	2	0.2183	14
Car17	0.1260	1	0.2729	3.5
Car18	0.0968	9	0.2811	2
Car19	0.1117	5	0.2236	10.5
Car20	0.0948	12	0.2519	6

nine deciles and becomes negative (with a value of −0.06) if the reduction is to the sample of three quartiles.

References

Barzilai, J., Cook, W. D., & Golany, B. (1987). Consistent weights for judgements matrices of the relative importance of alternatives. *Operations Research Letters, 6*(1), 131–134.

Brugha, C. M. (2000). Relative measurement and the power function. *European Journal of Operational Research, 121*, 627–640.

Likert, R. (1932). A technique for the measurement of attitudes. *Archives of Psychology, 140*, 1–55.

Roy, B. (2005). Paradigms and challenges. In J. Greco & S. Ehrgott (Eds.), *Multicriteria decision analysis: state of the art survey* (pp. 3–24). Boston: Springer.

Tryantaphilou, E., Lootsma, F. A., Pardalos, P. M., & Mann, S. H. (1994). On the evaluation and application of different scales for quantifying pairwise comparisons in fuzzy sets. *Journal of Multicriteria Decision Analysis, 3*, 133–155.

Chapter 5
Composition by Joint Probabilities

Abstract The difficulty involved in the composition of the criteria by weighted average, due to the difficulty of assigning weights to criteria and to sets of criteria, was already signaled in the preceding chapters. The probabilistic approach allows forgo the weighting of the criteria and get overall scores of preference by calculating joint probabilities.

Keywords Being the best · Being the worst · Conservative · Progressive · Optimist · Pessimist · All the criteria · At least one criterion

5.1 Approaches by Joint Probabilities

The choice between possible forms of composition of preferences in terms of joint probabilities is facilitated by its probabilistic character. For example, the alternatives can be ordered according to the probability of being the best by all the criteria simultaneously. To calculate this probability, assuming that the random components involved in the application of the different criteria are independent, it is enough to multiply the probabilities of being the best according to each criterion.

The form to obtain joint probabilities of preference depends on the point of view taken by the decision maker. Four forms of composition can be obtained according to the position taken by the decision maker on two questions, the first about conservatism and the second about pessimism. It is much easier to identify which of these views should guide the decision than choosing a set of weights to the criteria.

In the first aspect, the decision maker in the progressive side pays attention to maximize the probabilities of preference according to the criteria examined, while for that in the conservative side the odds that matter are of not minimizing such preferences. The progressive decision maker pays attention to the differences between the alternatives near the frontier of excellence; for the conservative one what is important is to differentiate those near the frontier of worst performance.

A. Parracho Sant'Anna, *Probabilistic Composition of Preferences, Theory and Applications*, Decision Engineering,
DOI 10.1007/978-3-319-11277-0_5

The term conservative, in this terminology, is associated to the idea of avoiding a losses, of concern on avoiding the negative extremes. On the other hand, the term progressive is associated to the desire to achieve the best gains.

With respect to optimism and pessimism, in the optimistic point of view it is enough to reach satisfactory results on at least one criterion. All the criteria are taken into account, but the composition uses the connective "or". What will be computed are the probabilities of maximizing (or not minimizing, if the view on the other axis is conservative) preference according to at least one among the multiple criteria. From the pessimistic point of view, the preference will be measured by the probability of maximizing (or not minimizing) the preference according to all the criteria.

So, four kinds of composition without the use of weights may be considered: (1) by the probability of being the best according to all the criteria considered, (2) by the probability of being the best according to at least one of the criteria, (3) by the probability of not being the worst by all the criteria considered, (4) the probability of not being the worst by at least one of the criteria.

In the progressive and pessimistic composition is used the joint probability of the intersection with respect to all the criteria of the events corresponding to the alternative receiving the best evaluation.

In the progressive and optimistic composition, the global preference score is given by the probability of the complement of the intersection of the events corresponding to not achieving the best evaluation.

In the conservative and pessimistic composition is used the joint probability corresponding to the intersection of the events of not receiving the worst rating.

In the conservative and optimistic composition is used the probability of the complement of the intersection of events corresponding to receiving the worst ratings.

By dividing the criteria into groups, these four joint probabilities can be computed for subsets of the set of criteria. After that, a choice between the optimistic and the pessimistic approaches will be applied to determine how to combine the groups to derive from the probabilities of preference inside each group a general score of preference.

The criteria may be divided, for example, into a group related to attributes of highest importance and another associated with secondary attributes, or into a group of inputs and other of outputs. The composition of the criteria of high priority would be taken from a pessimistic point of view, while secondary criteria would be seen from an optimistic point of view. The outputs would be treated from a progressive perspective, while inputs suggest a conservative treatment. In both cases, as the two groups cannot be thought as replaceable, a pessimistic approach must be taken to combine them and the final score will be given by the product of the probabilities of preference according to each group.

For instance, if the criteria are ranked as in the Analytic Network Process (ANP) of Saaty (1996), on criteria of benefits, opportunities, costs and risks, or of strengths, opportunities, weaknesses and threats, of the SWOT analysis of Humphrey (2005), for the first group, given its priority and positive feature, a progressive and pessimistic approach would apply. Opportunities, with a lower priority, would receive a progressive and optimistic approach. Similarly, to costs or weaknesses a conservative and pessimistic approach would be applied while to risks or threats a conservative but optimistic approach might be applied.

As a general rule, to compose the groups the product will be employed and for the rule of composition inside the groups, whenever the goal is to choose the best alternative, the progressive and pessimistic approach is what is needed. First, because the progressive approach focuses on the alternatives most likely to be chosen according to each criterion separately and it is natural that these are the alternatives more carefully measured. And the pessimistic approach leads to give more importance to good performances according to the highest number of criteria, while the optimistic presents a highest chance of good performances on isolated criteria determining the outcome.

Even if the decision maker is not able to choose a point of view to decide on the basis of joint probabilities, computing the preferences by every point of view and comparing the results of applying such different points of view can be employed to help identifying isolated criteria that favor certain alternatives, guiding a possible review of the application of the criteria, in a subsequent step.

5.2 Different Assumptions About Correlation

Another aspect to consider is the modeling of the correlation between disturbances affecting the evaluations according to different criteria.

If such correlation can be quantified, it can be introduced in the computation of the joint probabilities. However, quantifying correlations is difficult. For instance, in the choice of more efficient production units, correlation between the criteria may be expected, as the disturbances affecting the quantities of inputs used in the production process come to affect the volumes of product. But, how does one quantify the reciprocal influence between errors in the measurement of inputs and outputs?

The presence of correlation should be quantified only in the case of positive dependence. Negative correlation might also be considered in the case of contradictory features, but the choice between satisfying opposite preference motivations should be solved in the modeling of the criteria before entering the modeling of the disturbances affecting them.

In the absence of a means to directly calculate the correlations, an approach that can be taken is to compute the joint probabilities under hypotheses of independence and of maximal dependence between all the events of the same type, i.e., between all the events of the type 'maximizing preference according to different criteria', between all the events of the type 'not maximizing', between all the events of the type 'minimizing' and between all the events of the type 'not minimizing'.

If the difference between the scores obtained assuming independence and assuming maximal dependence reaches significant levels, research to determine the effect of intermediary levels of dependence must be undertaken.

In case of a lack of strong reasons to consider correlation, the assumption of independence should be preferred. This happens because, while the hypothesis of independence leads to calculating the probability of the intersection using the product of the probabilities of the events that are intersected, the hypothesis of maximal dependence, because the probability of the intersection cannot be larger than the probability of the events intersected, leads to calculating this probability using the minimum of the probabilities of such events. Thus, the assumption of independence, by involving the computation of the product of the exact values of the factors, employs more of the available information.

It is interesting to notice here that the composition by the minimum is equivalent to the application of the principle of necessity in Fuzzy Logic (Zadeh 1978).

Moreover, the hypothesis of maximal dependence provides a basis to establish a relation between the composition by joint probabilities and that by integration, as the combination by the minimum is equivalent to the application of a Choquet (1953) integral with respect to the capacity μ with μ(A) = 1 for every unitary set A.

When the number of criteria is large, the joint probability under the assumption of independence will present small values as the product of a large number of factors between zero and one. Under the assumption of maximal dependence, determining the joint probability using the smallest of the probabilities according to each criterion, the result will not be so small. However, simple procedures may be employed to avoid small final scores. It is enough to substitute for the product the geometric mean, or to standardize the final probabilities to sum one, dividing each product by the sum of all of them.

To illuminate the effect of the assumptions on dependence, the following formulae show the result of application to the different points of view of the extreme hypotheses of maximal dependence and independence. Here, M_j and m_j denote, respectively, the probability of the alternative maximizing and minimizing the preference according to the j-th criterion, π denotes the product, *min* denotes the minimum and *max* denotes the maximum.

5.2.1 Progressive and Pessimistic Point of View

$$\text{independence: } \pi_j\, M_j;$$

$$\text{maximal dependence: } min_j M_j.$$

5.2.2 Progressive and Optimistic Point of View

$$\text{independence: } 1 - \pi_j\left(1 - M_j\right);$$

$$\text{maximal dependence: } max_j M_j.$$

These formulae follow from the fact that the event of being preferred according to at least one criterion is the complement of the intersection of the events of not being preferred and, in the case of maximum dependence,

$$1 - \text{min}_j(1 - M_j) = 1 - (1 - \text{max}_j M_j) = \text{max}_j M_j.$$

5.2.3 Conservative and Pessimistic Point of View

$$\text{independence: } \pi_j(1 - m_j);$$

$$\text{maximal dependence: } 1 - max_j m_j.$$

Notice that there are two equivalent forms of computation under the assumption of maximal dependence, as the intersection of the probabilities of not being the worst can be given by $min_j(1 - m_j)$, as well as by $1 - max_j m_j$.

5.2.4 Conservative and Optimistic Point of View

$$\text{independence: } 1 - \pi_j m_j;$$

$$\text{maximal dependence: } 1 - min_j m_j.$$

This follows from the fact that not being the worst by at least one criterion is the complement of being the worst by all of them.

To compare in a more practical context the effect of the two assumptions of independence and maximal dependence, let us formally represent the application of them to the model derived from ANP. Denoting by P_{aj} the probability of preference by the a-th alternative according to the j-th criterion and indexing by b, o, c and r, respectively, the criteria of the four types, benefits, opportunities, costs and risks, the joint probability of preference by the a-th alternative would be given, assuming independence between events of maximizing preference according to different criteria, by:

$$\pi_b P_{ab}(1 - \pi_o(1 - P_{ao}))\pi_c(1 - P_{ac})(1 - \pi_r P_{ar})$$

and, assuming maximal positive dependence, by

$$min_b P_{ab} max_o P_{ao} min_c(1 - P_{ac})(1 - min_r P_{ar}),$$

for b varying along all benefits, o varying along all opportunities, c varying along all costs and r varying along all risks.

Another possible way to address the case of dependence is by conditional reasoning. After somehow ordering the criteria, measures of preference for each alternative can be calculated successively according to each criterion conditionally on the preference according to the previously applied criteria. This calculation may follow the same pattern of calculation of probabilities of being the best according to the isolated criterion by randomizing ranks. The conditional probabilities of preference thus obtained can be combined with the previous vector of preferences to derive a new joint probability of preference and so on until a joint preference by all the criteria is obtained.

5.3 Examples

Table 5.1 presents the results of the composition of the preferences for car models treated in the preceding chapters considering now the criteria divided into two

Table 5.1 Preferences assuming independence, priority to reliability and safety

Models	Scores	Standardized Scores	Ranks	Weighted average ranks
Car1	0.00003238	0.0032	15	13
Car2	0.00003055	0.0030	16	9
Car3	0.00086102	0.0859	5	3
Car4	0.00081111	0.0810	7	7
Car5	0.00002351	0.0023	19	20
Car6	0.00002218	0.0022	20	19
Car7	0.00083387	0.0832	6	6
Car8	0.00002417	0.0024	18	18
Car9	0.00054902	0.0548	13	12
Car10	0.00089604	0.0894	3	4
Car11	0.00002511	0.0025	17	17
Car12	0.00086922	0.0868	4	8
Car13	0.00058797	0.0587	12	16
Car14	0.00061253	0.0611	10	15
Car15	0.00003367	0.0034	14	11
Car16	0.00091023	0.0908	2	2
Car17	0.00093219	0.0930	1	1
Car18	0.00060376	0.0603	11	10
Car19	0.00069493	0.0694	8	5
Car20	0.00066615	0.0665	9	14

groups. Reliability and safety form a group of two criteria of high priority and the other five criteria are considered of lower priority.

The criteria of high priority are combined by a progressive and pessimistic point of view and those of lower of priority are combined by a progressive and optimistic point of view. The two groups are finally combined pessimistically, so that the final score is given by the probability of being preferred by the two groups.

Thus, the five criteria of low priority are combined in terms of probability of being the best by at least one of them and, assuming independence, the final score is obtained multiplying this joint probability by the product of the probabilities of preference by the two other criteria.

Beside the joint probability scores, is presented in Table 5.1 a column with the vector of scores standardized to sum 1 and a column with the ranks. For comparison, in a last column are repeated the ranks obtained by weighted average with the weights of Table 2.3.

It can be seen in Table 5.1 that the ranking obtained by joint probabilities is similar to that obtained by weighted average with higher weights for the criteria receiving now high priority, with the final choice of cars 17 and 16. The concordance between the results of application of these two approaches is confirmed by a ranks correlation coefficient of 0.88.

The joint probabilities in Table 5.1 are calculated assuming independence. The results of analogous computation assuming the hypothesis of maximal dependencies are presented in Table 5.2. It can be noticed how the hypothesis of maximal dependence considerably reduces the variation. In fact, only four different values are observed for the final scores there.

In this example of car models, as the evaluations by each criterion present only two possible values, minimizing is equivalent to not maximizing. The next example has a larger set of possible values.

In this second example are compared the performances of drivers of a fleet of urban buses. The drivers are grouped by the shifts of the bus line they work for. Four criteria are employed to assess individual performance: based on two process attributes and on two output attributes.

The process attributes are related to the speed kept in various parts of the route to avoid large spacing between two buses of the same line, which would increase the chance of passengers taking buses of competing companies. To build these process attributes, the times each vehicle passes by previously determined points are recorded. These attributes, denoted T_1 and T_2, are, respectively, the number of passages at each of these points before a predetermined small time threshold and after another higher one.

The output attributes, denoted by P_1 and P_2, are the number of passengers transported by the driver and by the whole shift of the driver.

Table 5.3 shows the weekly improvement in each of these attributes referring to ten drivers of the three shifts of a bus line. The attributes are measured in terms of weekly variation, what means, for instance, that a positive value of 3 in T_1 signifies that the driver presented this week less 3 transgressions of the passage thresholds relatively to the previous week.

Table 5.2 Preferences assuming dependence, priority to reliability and safety

Models	Scores	Standardized scores	Ranks
Car1	0.0585	0.1289	4
Car2	0.0016	0.0035	19
Car3	0.004	0.0088	12
Car4	0.004	0.0088	12
Car5	0.004	0.0088	12
Car6	0.0016	0.0035	19
Car7	0.0585	0.1289	4
Car8	0.0585	0.1289	4
Car9	0.004	0.0088	12
Car10	0.004	0.0088	12
Car11	0.0016	0.0035	19
Car12	0.0585	0.1289	4
Car13	0.004	0.0088	12
Car14	0.0585	0.1289	4
Car15	0.0585	0.1289	4
Car16	0.004	0.0088	12
Car17	0.0034	0.0075	17
Car18	0.004	0.0088	12
Car19	0.004	0.0088	12
Car20	0.0585	0.1289	4

Table 5.3 Bus Drivers Data

	T_1	T_2	P_1	P_2
D_{11}	0	−2	5	3
D_{21}	−3	−1	2	3
D_{31}	0	0	−4	3
D_{12}	0	0	4	9
D_{22}	0	0	3	9
D_{32}	0	0	2	9
D_{13}	−1	−1	4	8
D_{23}	−1	−3	3	8
D_{33}	−2	−4	1	8
D_{43}	0	0	0	8

The first two shifts, serving morning and afternoon periods of the day, have three buses, while the third, the evening shift, has four. D_{as} names the a-th driver of the s-th shift.

A first exploratory analysis of this data set highlights the best weekly evolution of the performance of the drivers of Shift 2, regarding both the final output and the process attributes.

Tables 5.4 and 5.5 show the result of the transformation into probabilities of reaching the upper and lower ranks in each attribute, assuming triangular distributions with extremes given by the lowest and highest values observed in Table 5.3, that means, −3 and 0 for T_1, −4 and 0 for T_2, −4 and 5 for P_1 and 3 and 9 for P_2. Very similar results are obtained if larger ranges are assumed for these distributions.

In Table 5.5, the high values—above 0.3—for the probabilities of driver D_{21} minimizing T_1, driver D_{33} minimizing T_2 and driver D_{31} minimizing P_1 deserve attention. This is an illustration of the ability of the probabilistic transformation to detach the extreme alternatives by considering the two frontiers, of best and worst performances. Looking only at Table 5.4 the values of the maximization

Table 5.4 Probabilities of being the best

	T_1	T_2	P_1	P_2
D_{11}	0.1532	0.0373	0.2606	0.0140
D_{21}	0.0092	0.0574	0.0632	0.0140
D_{31}	0.1532	0.1159	0.0200	0.0140
D_{12}	0.1532	0.1159	0.1594	0.2045
D_{22}	0.1532	0.1159	0.0955	0.2045
D_{32}	0.1532	0.1159	0.0632	0.2045
D_{13}	0.0289	0.2763	0.1594	0.0861
D_{23}	0.0289	0.0277	0.0955	0.0861
D_{33}	0.0140	0.0220	0.0465	0.0861
D_{43}	0.1532	0.1159	0.0367	0.0861

Table 5.5 Probabilities of being the worst

	T_1	T_2	P_1	P_2
D_{11}	0.0472	0.1013	0.0449	0.2691
D_{21}	0.4146	0.0648	0.0693	0.2691
D_{31}	0.0472	0.0475	0.4038	0.2691
D_{12}	0.0472	0.0475	0.0508	0.0246
D_{22}	0.0472	0.0475	0.0586	0.0246
D_{32}	0.0472	0.0475	0.0693	0.0246
D_{13}	0.0731	0.0376	0.0508	0.0298
D_{23}	0.0731	0.1922	0.0586	0.0298
D_{33}	0.1563	0.3664	0.0848	0.0298
D_{43}	0.0472	0.0475	0.1090	0.0298

Table 5.6 Scores for drivers by maximal dependence between outputs

	PP	OP	PC	OC
D_{11}	0.00013609	0.38264182	0.62736682	0.9994297
D_{21}	0.00000814	0.12933471	0.39821755	0.99879157
D_{31}	0.00003245	0.1817541	0.41506386	0.99487113
D_{12}	0.00377803	0.43374713	0.88187182	0.99994102
D_{22}	0.0022635	0.39070221	0.87462509	0.99993196
D_{32}	0.00149794	0.36894398	0.86468405	0.99991953
D_{13}	0.00039663	0.44403605	0.85359504	0.99994308
D_{23}	0.00022777	0.19726684	0.73780113	0.99987235
D_{33}	0.00005605	0.14776723	0.56259057	0.99960502
D_{43}	0.00036623	0.25451126	0.82338691	0.99984669

Table 5.7 Scores for drivers by independence

	PP	OP	PC	OC
D_{11}	0.00002085	0.40566928	0.59775511	0.99994223
D_{21}	0.00000047	0.13734483	0.37241305	0.99949898
D_{31}	0.00000497	0.2765888	0.39547284	0.99975638
D_{12}	0.00057879	0.49937583	0.84024747	0.9999972
D_{22}	0.00034677	0.46131982	0.83334278	0.99999677
D_{32}	0.00022948	0.44208337	0.82387097	0.99999618
D_{13}	0.00010959	0.4601034	0.82149987	0.99999584
D_{23}	0.00000658	0.21950255	0.68386786	0.99997547
D_{33}	0.00000123	0.15969849	0.47465767	0.99985528
D_{43}	0.00005611	0.34091341	0.78452305	0.99999272

probabilities for the other drivers would not call so much attention due to the ties in the frontier.

Tables 5.6 and 5.7 present the probabilistic scores obtained by joint probabilities, for the composition under different points of view: PP stands for pessimistic and progressive, OP for optimistic and progressive, PC for pessimistic and conservative and OC for optimistic and conservative.

Thus, for instance, in Table 5.6, the a-th entry of the column headed by PP presents the product of the minimum between the entries corresponding to T_1 and T_2 in the a-th line of Table 5.4 by the minimum between the other two entries of the same row, while in Table 5.7 it presents the product of the 4 entries of the a-th row of Table 5.4.

Analogously, the a-th entry of the column headed by PC presents, in Table 5.6, the difference to 1 of the product of the minimum between the differences to 1 of each of the entries corresponding to T_1 and T_2 in the a-th row of Table 5.4 multiplied by the product of the differences to 1 of each of the other two entries of the a-th row of that Table, while in Table 5.7 it presents the difference to 1 of the product of the differences to 1 of the four entries corresponding to T1, T2, P1 and P2 in that same row.

If the goal of the evaluation is to determine the values of bonuses to be paid to motivate drivers to increase output, a progressive point of view may be the most appealing. Besides, a pessimistic point of view may be more appropriate to make the drivers pay attention to both output and process attributes. This approach may nevertheless be sometimes replaced by an optimistic approach that would make easier to the drivers improve performance by focusing attention on a simple goal.

In the present study these distinctions may become irrelevant as high correlations can be noticed between all the vectors of final scores.

An even stronger concordance can yet be found between the scores in Tables 5.6 and 5.7 for the same composition point of view. This demonstrates the resistance of the composition assumptions to the effects of dependence. In fact, changing the assumption from maximal dependence to independence would result in significant change only in the ranks of D_{13} for the optimistic approaches.

This last change is due to the greater importance given to the outputs if independence is assumed. To avoid giving higher weight to a group with a larger number of independent criteria, the probabilistic composition may be applied first within the groups, replacing the products of the probabilities of preference by criteria in the same group by geometric means.

In the present example, slightly different results are obtained if this procedure is adopted. For the scores for the group of two process criteria, assumed as independent, computed using square roots of the products while the scores for the group of output criteria, assumed dependent, composed using the minimum, the final scores in Table 5.8 are obtained. Comparing the ranks derived from Table 5.8 to those from the previous tables, again the few changes observed are small.

Table 5.8 Scores for equal group weights and dependence

	PP	OP	PC	OC
D_{11}	0.0023	0.2770	0.7509	0.9948
D_{21}	0.0003	0.0941	0.4828	0.9912
D_{31}	0.0019	0.1676	0.6288	0.9844
D_{12}	0.0209	0.3075	0.9165	0.9983
D_{22}	0.0162	0.2817	0.9127	0.9982
D_{32}	0.0132	0.2690	0.9075	0.9981
D_{13}	0.0034	0.3657	0.8895	0.9985
D_{23}	0.0025	0.1171	0.7720	0.9969
D_{33}	0.0009	0.0870	0.5970	0.9921
D_{43}	0.0065	0.2055	0.8856	0.9973

References

Choquet, G. (1953). Theory of capacities. *Annales de l'Institut Fourier, 5*, 131–295.
Humphrey, A. S. (2005). *SWOT Analysis for Management Consulting*. SRI Alumni Newsletter, December, pp. 7–8.
Saaty, T. L. (1996). *Decision making with dependence and feedback: The analytic network process*. Pittsburgh: PA, RWS Publications.
Zadeh, L. A. (1978). Fuzzy sets as the basis for a theory of possibility. *Fuzzy Sets and Systems, 1*, 3–28.

Chapter 6
Composition by DEA Distance to the Frontier

Abstract In the preceding chapter the composition of the preferences according to the different criteria was made by rules that do not weight the criteria. These rules were based on different points of view formulated in terms of joint probabilities. A different approach to avoid a previous determination of weights, based on multivariate distances to the frontiers, is considered in this chapter.

Keywords Data envelopment analysis · Frontier · Human development index

6.1 Data Envelopment Analysis

Data Envelopment Analysis (DEA) emerged in the late 1970s, with the work of Charnes et al. (1978). The basic idea behind DEA consists of comparing the alternatives corresponding to production units by evaluating their efficiency in terms of ratios between linear combinations of the values of outputs and inputs, with coefficients varying freely.

If there is only one product from a unique resource, the efficiency is naturally measured by the ratio between the amount of output produced and the amount of resource employed to produce it. In the case of multiple inputs combined to produce different amounts of different products, a similar ratio can be employed if it is possible, by means of unitary prices or any other type of coefficients, to obtain a unique aggregate measure of output produced and another unique aggregate measure of input employed.

DEA is employed to measure the efficiency when there is freedom of choice of coefficients to combine outputs and inputs. The efficiency of a production unit is then measured by the distance between the ratio obtained applying to its outputs and inputs the most favorable coefficients and the highest ratio that would be obtained applying the same coefficients to outputs and inputs observed in some competing alternative.

A mathematical programming algorithm is employed to determine this optimal distance from the productivity ratio associated with the alternative under evaluation to the productivity ratios in the frontier of highest productivities. An alternative is considered efficient and receives an efficiency score of 1 if its vector of inputs and

A. Parracho Sant'Anna, *Probabilistic Composition of Preferences, Theory and Applications*, Decision Engineering,
DOI 10.1007/978-3-319-11277-0_6

outputs belongs to this excellence frontier. Otherwise, it is said to be inefficient and receives a score of relative efficiency determined by the quotient between such productivity ratios.

There are two classic models in DEA. The first is called CCR, in reference to Charnes et al. (1978). The second, called BCC, was proposed by Banker et al. (1984). Later successive developments led to the construction of a series of DEA models, considering, for instance, restrictions to the vector of weights, as well as randomness. However, when establishing a setup to combine probabilistic preferences, it is convenient to not deviate from a fixed pattern. For this reason, only the two general optimization procedures of the CCR and BCC models will be employed here in the probabilistic composition of criteria.

In both these procedures, the distance from the vector of values of the inputs and outputs in the alternative being evaluated to a piecewise linear frontier generated by linearly combining observed inputs of other alternatives is minimized. In the BCC approach, the coefficients of the linear combinations that form the frontier are forced to sum to 1. In the CCR approach, they can present a sum smaller than 1.

This flexibility in the sum of coefficients of the CCR model corresponds to allowing for introducing in the composition of the frontier an alternative of null inputs and null outputs. This is equivalent to allow for reducing the use of inputs and production of outputs in any unit in the frontier on a proportional basis, as if returns to scale were constant.

In the BCC model, the frontier of excellence is formed only by points intermediary between those representing observed alternatives. That means that the scores of efficiency of the alternatives cannot be reduced by comparison to fictitious production units of smaller dimensions.

The efficiency of each alternative is calculated by solving an optimization problem proper for that alternative. The CCR optimization problem has as objective function to maximize the quotient between the efficiency ratio of that alternative and the best efficiency ratio with the same multipliers in the entire set of alternatives. By fixing the value 1 for the linear combination of the inputs that constitutes the denominator of the ratio, this fractionary optimization problem can be formulated in the following linear form:

$$\text{Maximize } \theta_{ao} = \sum_r \mu_r O_{rao}$$

subject to

$$\sum_s v_s I s_{ao} = 1,$$

$$\sum_r \mu_r O_{ra} - \sum_a v_s I_{sa} \leq 0, \quad \text{for every alternative a, from 1 to n}$$

$$\mu_r > 0 \text{ and } v_s > 0 \text{ for all r and s,}$$

for r varying from 1 to m_1 and s varying from 1 to m_2, m_1 denoting the number of outputs and m_2 denoting the number of inputs, O_{ra} denoting the value of the r-th output at the a-th alternative, I_{si} denoting the value of the s-th input at the a-th alternative and i_o denoting the alternative being evaluated.

This problem is equivalent to its dual linear programming problem of minimizing the fraction by which the values of the inputs of the production unit evaluated can be reduced in such a way as to keep its output/input ratio lower than that obtained from any combination of all the alternatives.

The score for the a_o-th alternative is the fraction θ_{ao} such that, for some set of coefficients $\lambda_1 \ldots \lambda_n$, for every input, the linear combination with these coefficients of the values of the input in the n alternatives is smaller than the fraction of the value of that input for the a_o-th alternative, while the linear combination with these same coefficients of the n values of every output is larger than the value of that output for that a_o-th alternative. This dual problem may be formulated in precise terms, with the same notation, as:

$$\text{Minimize } \theta_{ao}$$

subject to

$$\sum_a \lambda_a I_{si} - \theta_{ao} I_{sao} \leq 0, \quad \text{for all s from 1 to } m_1,$$

$$O_{rao} - \sum_a \lambda_a O_{ra} \leq 0, \quad \text{for all r from 1 to } m_2,$$

$$\lambda_a \leq 0, \quad \text{for all a from 1 to n.}$$

To avoid the hypothesis of constant returns to scale, in the BCC model the constraint $\Sigma_a \lambda_a = 1$ is added to this formulation.

The CCR model is completely symmetric with respect to fixing an upper bound for inputs and maximizing outputs or fixing a lower bound for outputs and minimizing inputs, in the sense that, by taking this second approach, the score given by the lower bound obtained is precisely the inverse $1/\theta_{ao}$ of the upper bound θ_{ao} determined by solving the problem presented above.

In the BCC model, there is no such precise relation. Two different vectors of scores are then obtained according to minimization of inputs or maximization of outputs.

Thus, two more algorithms are available: the BCC model oriented to the minimization of inputs and the BCC model oriented to the maximization of outputs.

6.1.1 BCC Model Oriented to the Minimization of Inputs

$$\text{Minimize } \theta_{ao}$$

subject to

$$\sum_a \lambda_a I_{sa} - \theta_{ao} I_{sao} \leq 0, \quad \text{for all s from 1 to } m_1,$$

$$O_{rao} - \sum_a \lambda_a O_{ra} \leq 0, \quad \text{for all r from 1 to } m_2,$$

$$\sum_a \lambda_a = 1,$$

$$\lambda_a \geq 0, \quad \text{for all a from 1 to n.}$$

6.1.2 BCC Model Oriented to the Maximization of Outputs

$$\text{Maximize } \theta_{ao}$$

subject to

$$\theta_{ao} O_{rao} - \sum_a \lambda_a O_{ra} \leq 0, \quad \text{for all r from 1 to } m_2,$$

$$\sum_a \lambda_a I_{sa} - I_{sao} \leq 0, \quad \text{for all s from 1 to } m_1,$$

$$\sum_a \lambda_a = 1,$$

$$\lambda_a \geq 0, \quad \text{for all a from 1 to n.}$$

For multiple criteria composition, the more natural mode of use of the DEA approach is to measure all the criteria in an increasing form and treat them as outputs generated by a constant input. With constant input, there is no reason to consider variable returns to scale, and the model takes the simpler form:

$$\text{Maximize } \theta_{ao} = \sum_r \mu_r O_{rao}$$

subject to

$$\sum_r \mu_r O_{ra} \leq 1, \quad \text{for every alternative a, from 1 to n,}$$

$\mu_r > 0$ for all r, r varying from 1 to m, the number of criteria, O_{ra} denoting the evaluation of the a-th alternative by the r-th criterion and a_0 denoting the alternative being evaluated.

6.2 Use of DEA Scores

The composition of probabilities of preference according to multiple criteria may employ the DEA multidimensional distances to the frontiers of best or worst performance. The scores of efficiency generated by the algorithms of DEA take the role of global scores of preference. DEA models are based on efficiency ratios between input and output variables, but they can be applied to any situation by treating all criteria as outputs generated by an input of identical values along all the alternatives.

This approach may be applied even in the problem of comparing production units. In that case, the output/input productivity ratios may be treated as outputs of constant input.

Changing signs if necessary, the evaluations according to every criterion can be oriented in such a way that higher evaluations correspond to better alternatives. With this, a DEA model with a constant input and with the probabilities of maximizing preference according to each criterion as outputs can always be employed.

DEA scores do not depend on the scales of measurement and do not involve weighting variables. Nevertheless, they are heavily affected by outliers. Any alternative with an extreme low value in only one input or an extreme high value in only one output will be necessarily evaluated as fully efficient. Its extreme value will also strongly affect the score of inefficient units.

By taking into account distances to other alternatives besides those in the frontier, the probabilistic transformation into probabilities of being the best increases the resistance of the scores of the alternatives outside the frontier against the influence of such outliers. However, if the DEA algorithm is used, even with the probabilistic transformation, care must be taken to detect the possibility of an extreme value in some variable leading to an efficiency score equal to 1.

In addition to the model with constant input, the probabilistic composition can also use DEA algorithms to evaluate efficiency of production units employing combinations of inputs to generate sets of outputs. The probabilistic approach to this problem can take as criteria the output/input ratios for the different pairs of input and output or take each input and each output as individual criteria.

In this latter form of modeling, the individual probabilities of preference will be the probabilities of maximizing some measure of revenue from the sale of each output and the probabilities of minimizing some measure of cost of acquisition of each input. In the other form, which takes as criteria the output/input ratios, the final scores will measure efficiency in a manner more similar to that traditionally used in DEA.

Preference scores for the cars in the example models in the preceding chapters can be obtained by employing DEA algorithms to combine the probabilities of preference according to the different criteria. As there are not causal relationships between the attributes considered in this model, a constant inputs DEA model with one output for each attribute must then be used. However, the result of the application of the DEA algorithm will present as fully efficient, i.e., with a global score of preference of 1, all the alternatives. This follows from the fact that, with a large enough number of criteria, it becomes possible to find for each alternative one criterion for which the alternative is better than all the others, which is enough to guarantee full efficiency.

A similar situation is that of the bus schedules example. Because the goal intended in that case is raising quality and not productivity in the use of any inputs, the process variables T_1 and T_2 should be maximized in the frontier of excellence, as well as the outcome variables. This may be dealt with again inside the DEA framework by applying the constant inputs approach.

In that case, a majority of the alternatives will again be fully efficient. It is easy to see, for instance, that the drivers of the second shift, which maximizes the last output, will be in the frontier and, consequently, have an efficiency score equal to 1.

It may be noticed anyway, even in the simple examples above, that the constancy of the results is limited to the conceptual approach taken. It would be confirmed if, for instance, criteria to globally evaluate the performance of each shift with respect to the process variables already in the analysis were added. If, on the contrary, the change extends beyond the scope of this analysis to take into account other variables, such as contribution to vehicle maintenance or careful driving, for instance, the results might be entirely altered. In that case, the simplicity of the vector of scores generated by the DEA approach may be an advantage.

6.3 A Human Development Index

This example shows how, even with the same criteria, the form of composition may be employed to stress different points of view. Here, the probabilistic composition is employed to offer diversified options of computation to the Human Development Index—HDI (UNDP 2014). This index was designed to provide a single counterpoint to the per capita Gross National Product (GNP) as a development measure, as this does not take into account important aspects of quality of life that do not depend on the volume of goods traded in the market.

Social values are complex and cannot be measured directly. Social indicators are objective measures of simple attributes that are supposed to be in some way correlated with the qualities that matter. Therefore, there is a risk in the use of social indicators. Used to set goals, they can divert initiatives from the true values to false representatives.

This is what happens with the gross national product or the national income. Even when computed on a per capita basis they deviate the focus of promoting social development to isolated economic goals. The HDI was created to try to reduce the importance attached to strictly economic indicators. In turn, to be a simple indicator with the properties of direct and objective communication, it also faces the same risk of moving away from the dimensions to be measured.

While, by one side, it is important to have a single indicator to substitute for per capita GNP, on the other side it is important to validate and refine this new indicator by comparing to other complex indicators in the same way as other economic indices are compared to per capita GNP.

In 2010 the form of computation of HDI was changed. From the viewpoint of the calculation algorithm, the main change is that, instead of a weighted arithmetic mean, a geometric mean is now employed. This approximates the algorithm to the joint probabilities approach for probabilistic composition.

Other measures using the same components of HDI, but using probabilistic composition to combine them into a single measure can be built with different properties. A feature that all the measures discussed here share is taking as their starting point the preliminary transformation of the partial indicators of preference into probabilities of reaching the frontier of best or of worst performance in each of them.

By 2009, the partial indicators combined in the HDI were:

Life expectancy at birth, as an indicator of development on the dimension of health;

Literacy, weighing 2/3, combined with the gross school enrolment rate (ratio between the total number of students attending school in the three levels of education and the total number of persons of school age), weighing 1/3, as indicator of development in the dimension of education;

Neperian logarithm of the per capita gross domestic product, as measured in terms of PPP—purchase power parity (World Bank 2014), as an indicator of economic welfare of the population of the country.

Each of these indicators was transformed to a scale from 0 to 1 by subtracting an absolute minimum and dividing by an absolute amplitude. The maximum and minimum for longevity in a country were then fixed in 25–85 years. For the per capita GDP, the extremes were 100 and 40,000.

Finally, the HDI was calculated as the arithmetic mean of the measures of development in the three directions.

Currently slightly different partial indicators and composition rule are employed. The three dimensions remain.

The health indicator is still life expectancy at birth, measured as before, but with limits of 20–83.57 years. As an indicator of income the Neperian logarithm of per capita GDP was replaced by that of per capita Gross National Income, with limits of 100 and 87478 PPP.

The biggest change occurred in the evaluation of development in the educational dimension, whose indicator is now the geometric mean between two measures: the average number of years that a person 25 years old now attended school and the expected value of the number of years that a person 6 years old now will attend school. This second parameter is estimated from the present distribution of school enrolment provided by censuses and national sample surveys and employing the estimation methodology of Barro and Lee (2010).

Before calculating their geometric mean, these two indicators are standardized to values between zero and one. The average number of school years of the adult observed in the country is divided by the observed maximum of 13.3 and the expected number of school years of the child by the allowed maximum of 18.

By using, instead of the arithmetic mean of the three indicators, the geometric mean, the analyst considers that the influence of the three factors accumulates in a multiplicative basis. Thinking of each of the partial indicators as a probability to meet a standard of welfare—which seems to be the real aim of building measures between zero and 1—one may think HDI as the joint probability of reaching excellence in the three dimensions.

Another interesting aspect is the application in the algorithm of the logarithmic transformation to the income value. The logarithmic transformation is used to reduce the importance of variations in the upper end of the indicator. A similar effect may be obtained by applying the transformation into the probability of reaching the lowest extreme.

In fact, the indicators developed below avoid two ad hoc transformations: the logarithmic transformation and the prior determination of absolute maximum and minimum for each component. The global indicator is constructed by a combination of the probabilities of reaching the frontier of worst performance in each dimension.

HDI was created to emphasize the shortcomings in the today prevailing concept of development. Employing the standardization in terms of probabilities of being the worst has the advantage of calling attention to where the unattended features of development are located by reducing the differences between the relatively satisfactory performances and extending the differences between those countries in need of larger improvement.

Table 6.1, presenting the probabilities of being the worst for 30 countries selected among those of best and worst performance in each component of the index, illustrates this feature of the probabilistic transformation. These scores were obtained by calculating the probability of each country being the worst in the entire population of 187 countries, employing the data of 2013 of the United Nations Development Programme (UNDP 2014). This data set is available at http://hdr.undp.org/en/reports.

The computation of the probabilities of being the worst is performed adopting triangular distributions with equal amplitudes for all countries and extreme extended by 10 %, for all the criteria.

It can be seen in Table 6.1 that the first 17 countries of the list have equal scores for longevity, if rounding to the third decimal place is employed. In the other

Table 6.1 Probabilities of minimization of the HDI components

Country	Longevity	Observed schooling	Expected schooling	Income
Switzerland	0.0026	0.0029	0.0034	0.0017
Australia	0.0026	0.0026	0.0026	0.0019
Israel	0.0026	0.0027	0.0034	0.0024
France	0.0026	0.0030	0.0033	0.0021
Sweden	0.0026	0.0027	0.0033	0.0019
Norway	0.0026	0.0025	0.0030	0.0014
Japan	0.0025	0.0027	0.0035	0.0020
Iceland	0.0026	0.0030	0.0028	0.0022
Netherlands	0.0027	0.0027	0.0031	0.0018
New Zealand	0.0027	0.0025	0.0026	0.0025
Germany	0.0027	0.0026	0.0032	0.0019
Hong Kong	0.0025	0.0032	0.0034	0.0015
United States	0.0028	0.0024	0.0031	0.0016
Singapore	0.0027	0.0029	0.0035	0.0018
Qatar	0.0029	0.0043	0.0047	0.0009
Liechtenstein	0.0028	0.0031	0.0048	0.0009
Kuwait	0.0032	0.0052	0.0038	0.0013
Mali	0.0134	0.0159	0.0096	0.0090
Equatorial Guinea	0.0144	0.0059	0.0088	0.0027
Chad	0.0185	0.0213	0.0098	0.0086
Burkina Faso	0.0086	0.0246	0.0111	0.0087
Mozambique	0.0160	0.0266	0.0069	0.0089
Central African Republic	0.0218	0.0091	0.0114	0.0091
D. R. Congo	0.0239	0.0091	0.0078	0.0096
Sierra Leone	0.0279	0.0096	0.0100	0.0090
Papua New Guinea	0.0053	0.0081	0.0154	0.0077
Djibouti	0.0071	0.0083	0.0160	0.0077
Niger	0.0093	0.0228	0.0228	0.0091
Eritrea	0.0056	0.0093	0.0271	0.0093
Sudan	0.0057	0.0102	0.0288	0.0081

indicators, also, it can be seen that the evaluations of these 17 countries are much closer together than the others.

The results of the application of four different forms of probabilities composition to the data in Table 6.1 are shown in Table 6.2.

Table 6.2 HDI by different forms of composition

	Pessim.	Pessim.	Optim.	Optim.	DEA	DEA	Choquet	Choquet	UN
Country	Score	Rank	Score	Rank	Score	Rank	Score	Rank	Rank
Switzerland	0.9539	1	0.9767	11	0.9999	17	0.9974	10.5	9
Australia	0.9534	2	0.9752	17	1	5	0.9976	2	2
Israel	0.9533	3	0.9685	29.5	0.9999	17	0.9972	26.5	16
France	0.9531	4	0.9725	24	0.9999	17	0.9973	19.5	20
Sweden	0.953	5	0.9757	14	0.9999	17	0.9974	10.5	7.5
Norway	0.9527	6	0.9783	5	1	5	0.9977	1	1
Japan	0.9525	7	0.9743	21.5	1	5	0.9973	19.5	10
Iceland	0.9524	9	0.9715	25	1	5	0.9974	10.5	13.5
Netherlands	0.9522	11.5	0.9760	12	0.9999	17	0.9975	4	4
New Zealand	0.9522	11.5	0.9663	32	1	5	0.9974	10.5	6
Germany	0.952	15	0.9755	15.5	0.9999	17	0.9974	10.5	5
Hong Kong	0.9498	22	0.9777	8	1	5	0.9975	4	13.5
United States	0.9487	25	0.9773	9	1	5	0.9975	4	3
Singapore	0.9461	29	0.9789	3.5	0.9999	17	0.9974	10.5	18
Qatar	0.9196	72	0.9821	1	1	5	0.9972	26.5	36
Liechtenstein	0.9139	77.5	0.9819	2	1	5	0.9972	26.5	24
Kuwait	0.8966	94	0.9789	3.5	0.9999	17	0.9968	36.5	54
Mali	0.7171	165.5	0.8194	184	0.993	185.5	0.9872	174.5	182
Equatorial Guinea	0.7162	167	0.9626	40	0.9982	138	0.9914	157	136.5

(continued)

Table 6.2 (continued)

	Pessim.	Pessim.	Optim.	Optim.	DEA	DEA	Choquet	Choquet	UN
Chad	0.7048	171	0.8258	177	0.993	185.5	0.9839	186	184
Burkina Faso	0.7029	172	0.8248	178	0.9939	182	0.9860	179.5	183
Mozambique	0.7011	174	0.8201	182	0.9957	174.5	0.9828	187	185
C. African Rep.	0.6995	175.5	0.8176	185	0.9933	183	0.9859	181	180
D. R. Congo	0.697	179.5	0.8125	187	0.9948	180	0.9858	182.5	186.5
Sierra Leone	0.6936	182	0.8197	183	0.9929	187	0.9844	184	177
PapuaNewGuinea	0.6877	183	0.8852	159	0.9972	156	0.9905	167	156
Djibouti	0.6863	184	0.8453	172	0.9954	177	0.9897	170	164
Niger	0.6768	185	0.8173	186	0.9932	184	0.9863	178	186.5
Eritrea	0.6737	186	0.8750	163	0.9969	162	0.9860	179.5	181
Sudan	0.6728	187	0.8729	165	0.9968	165	0.9858	182.5	171

The first two scores are obtained by computing joint probabilities, assuming maximal dependence between the components, what means using for the probability of the intersection the lowest probability. The first of these scores follows from application of the pessimistic approach, i.e., evaluates the country by the probability of not being the worst by any criterion. The second results from the optimistic joint probability composition: assesses the country by the probability of not being the worst by at least one of the criteria.

The third form of composition is also optimistic, but uses the DEA algorithm for constant input. The score is given by the proximity to the DEA excellence frontier in the component more favorable to the country.

Finally, the fourth score is given by Choquet integral, modeling the preference among the criteria by a capacity with equal preference of 1/3 for each indicator, assuming additivity for different dimensions but assuming for the two indicators in the educational dimension, instead of additivity, full substitutability.

One can easily see in Table 6.2 the strong concordance between the evaluations by the different approaches, with each other and with the HDI presently computed by UNDP, whose ranks are in the last column of Table 6.2. If we consider the vectors of evaluations of all the 187 countries, the lowest Spearman correlation coefficient with the UN HDI ranking, obtained by optimistic joint probability, is 0.94. The others vectors of ranks present correlations with UN HDI ranks of 0.95 for DEA, 0.96 for the pessimistic composition and 0.99 for the composition by the Choquet Integral with respect to the capacity above referred. For 165 of the 187 countries, the difference between the ranks by the current model and by the Choquet integral is lower than 10. It is also 0.99 the rank correlation coefficient between the HDI as currently computed and the result of applying the geometric mean algorithm employed in it to the probabilities of reaching the extremes of worst performance substituting for the partial components.

This close agreement demonstrates the robustness of the approach, the final result not influenced by changes in the calculation. Nevertheless, it is possible to notice some systematic differences between the classifications. Considering the goal of taking into account all the dimensions considered, it is interesting to highlight the differences caused by the use of the pessimistic approach, which maximizes this feature.

Table 6.2 detaches, as examples of countries whose pessimistic score is improved by performing well with respect to all the criteria, Switzerland, Israel, France and Sweden, among those in the best positions. On the other hand, Singapore, Kuwait, Papua New Guinea and Djibouti are countries with worse ranks if the pessimistic composition approach is taken. The first two have good performance in terms of income, what is enough to place them in good position by some forms of composition, not accompanied by good performance in educational indicators. The other two, on the contrary, have poor performances with respect to income, contradicting a better performance for the other indicators, especially longevity.

A strategy that encourages countries to seek escape from underdevelopment in some special aspect may also be preferred, as advances in some dimension may have subsequent effects on the others. For such strategies, the optimistic joint probability approach and the DEA algorithm may be contemplated.

Differences between these two last approaches must also be noticed. The use of DEA corresponds to a more radical option for valorization of a marked advance in one of the indicators isolated. In the optimistic probabilistic approach, less significant advances combined in two different dimensions have a positive effect on the overall rating that does not appear in the composition by the DEA algorithm.

For example, countries like Liechtenstein and Qatar, which stand out for income, but do not have such a good behavior in the educational dimension, have a better evaluation when the efficiency in reaching DEA frontier is applied, but have a worse evaluation by the optimistic joint probabilistic composition, though not as bad as by the pessimistic composition.

It is also interesting to consider the cases of Japan and Iceland, which reach the DEA frontier and are rated in the pessimistic approach that rewards consistency of evaluations according to all criteria in a better rank than when evaluated by the optimistic joint probability composition. Under this approach they are overtaken by countries with scores close to the position of excellence in more than one criterion.

At the other extreme, call attention the positions of Eritrea and Sudan, with the last positions in the pessimistic ranking, but escaping that position in the optimistic probabilistic composition and in the assessment by the DEA algorithm, due to their longevity scores. On the other hand, Sierra Leone, Congo and Central African Republic, despite the worst performance by this criterion, have a relatively better evaluation under the pessimistic assessment, due to a relatively better position by other components.

The largest discrepancy between assessments for the same country in Table 6.2 is provided by Equatorial Guinea, a country for which the indicator of per capita income ensures the fortieth position in the optimistic probabilistic classification, while by the pessimistic classification, taking into account more strongly the proximity to the frontier of worst performance in the other indicators, is placed among the countries with the worst scores.

References

Banker, R., Charnes, A., & Cooper, C. (1984). Some models for estimating technical and scale inefficiencies in data envelopment analysis. *Management Science, 30*(9), 1078–1092.

Barro, R. J., & Lee, J. W. (2010). *A new data set of educational attainment in the world, 1950–2010*. National Bureau of Economic Research Working Paper 15902. Retrieved 2010 from http://www.nber.org/papers/w15902.

Charnes, A., Cooper, W., & Rhodes, E. (1978). Measuring the efficiency of decision making units. *European Journal of Operational Research, 2*(6), 429–444.

UNDP. (2014). Human Development Report. UNDP–HDRO, New York. Retrieved 2014 from http://hdr.undp.org/en/reports.

World Bank. (2014). PPP conversion factor (GDP) to market exchange rate ratio. Retrieved 2014 from http://data.worldbank.org/indicator/PA.NUS.PPPC.RF.

Chapter 7
Dynamic Probabilistic Indices

Abstract A DEA approach to access the evolution of productivity along time exploring the idea of the Malmquist index involves calculating the ratio between the distances to the production frontier of 2 years of evaluation of the alternative being evaluated: one representing the observed production during the year to which the frontier is built and the other representing the observed production of the next or the previous year. The same type of substitution is here applied with a probabilistic composition algorithm to generate probabilistic Malmquist indices.

Keywords Malmquist index · DEA Malmquist index · Probabilistic Malmquist index

7.1 Malmquist Indices

The Malmquist Index, as conceived by Malmquist (1953), aimed to assess the evolution of consumer behavior. Caves et al. (1982) proposed the use of the same idea in an index to assess the evolution of productivity. A DEA approach to the Malmquist index was developed by Fare et al. (1994).

The Malmquist DEA indices apply the algorithms of linear programming of DEA to the calculation of the Malmquist evolution of productivity. The basic concept involves calculating the ratio between the distances to the production frontier of 2 years of evaluation of the alternative being evaluated: the first representing the observed production during the year to which the frontier is built, and the other representing the observed production of the next or the previous year. The same type of substitution can be applied with a probabilistic composition algorithm instead of a DEA algorithm.

A. Parracho Sant'Anna, *Probabilistic Composition of Preferences, Theory and Applications*, Decision Engineering,
DOI 10.1007/978-3-319-11277-0_7

7.2 Variants for the Computation of Malmquist Indices

To neutralize the effect of technological change, the comparison to the frontier of excellence can be repeated, taking as reference, successively, the production frontier of the two consecutive years between which the progress of the production unit is being evaluated. This means realizing two computations: first, taking as reference the frontier of the first year and dividing the distance to this frontier (measured in terms of DEA efficiency score or of probability of being the best) of the vector formed with the production figures of the subsequent year by that distance of the vector formed with the production figures of the initial year; then, calculating the ratio between the distances of the same two vectors to the frontier of the second year. The Malmquist index most used is formed by the geometric mean between these two ratios.

Thus, to calculate the index, for each evaluated alternative, the computation of four measures of distance to the frontier is required. Using the notation d(a, t, u) to denote the distance of the vector of observations of the a-th alternative at instant t to the frontier formed using the observations of the various alternatives at instant u, these measures are:

(1) the distance to the production frontier of time t of the vector observed in the a-th alternative at time t, denoted by d(a, t, t);
(2) the distance to the production frontier of time t of the vector observed in the a-th alternative at time t + 1, denoted by d(a, t + 1, t);
(3) the distance to the production frontier of time t + 1 of the vector observed in the a-th alternative at time t, denoted by d(a, t, t + 1);
(4) the distance to the production frontier of time t + 1 of the vector observed in the a-th alternative at time t + 1, denoted by d(a, t + 1, t + 1);

By taking as reference only the frontier of time t, the index of productivity evolution is defined by d(a, t +1, t)/d(a, t, t). By sticking to the frontier of time t + 1, the index will be defined by d(a, t + 1, t + 1)/d(a, t, t + 1). To take into account the evolution of the frontier, the geometric mean of these two indices in the geometric mean Malmquist index is finally used:

$$[(\mathrm{d}(\mathrm{a},\mathrm{t}+1,\mathrm{t})/\mathrm{d}(\mathrm{a},\mathrm{t},\mathrm{t}))(\mathrm{d}(\mathrm{a},\mathrm{t}+1,t+1)/\mathrm{d}(\mathrm{a},\mathrm{t},\mathrm{t}+1))]^{1/2}.$$

Here, the distance to the frontier is given in probabilistic terms. Thus, denoting by P(a, t, u) the probabilistic evaluation of the a-th alternative when the values observed at time t are introduced into the analysis, together with all the values observed at time u, the Malmquist probabilistic rate of improvement is given by the square root of the product of the ratios $\mathrm{P}(\mathrm{a},\mathrm{t}+1,\mathrm{t})/\mathrm{P}(\mathrm{a},\mathrm{t},\mathrm{t})$ and $\mathrm{P}(\mathrm{a},\mathrm{t}+1,\mathrm{t}+1)/\mathrm{P}(\mathrm{a},\mathrm{t},\mathrm{t}+1)$:

$$M_P = \{[P(a, t+1, t)/P(a, t, t)]\,[P(a, t+1, t+1)/P(a, t, t+1)]\}^{1/2}.$$

It should be noticed that, by conceiving the distance to the frontier of excellence as a measure that is greater the farther the evaluated alternative is from the frontier, the increase in this value indicates a worsening in the productivity of the alternative. In the DEA-Malmquist index, the DEA efficiency score is used as a distance measure, which is greater the shorter the distance to the frontier of excellence; thus, the higher the value of the index is, the larger the improvement in the performance of the production unit evaluated.

In the probabilistic approach, we have the possibility of a choice between a progressive view, looking at distances to the frontier of excellence, and a conservative one, considering distances to the frontier of worst performance. In DEA, this last option corresponds to the inverted DEA concept proposed by Yamada et al. (1994).

Concomitant analysis of the same problem by the DEA and Inverted DEA models may indicate that an alternative is both efficient and inefficient according to the approach taken. Entani et al. (2002) classify this type of alternative as peculiar and suggest that the efficiency assessment be presented in terms of a range covering both results.

In the probabilistic approach, it is also possible to combine progressive and conservative evaluations in a compound Malmquist index.

By comparing to all other alternatives and not only to those at the frontier, the probabilistic composition scores are naturally less vulnerable to the influence of extreme values. The combination of scores relative to the two extremes provides even more stable results.

When applying either a DEA algorithm or a joint probability, one must be prepared to face the possibility of large variations from one year to the next in the set of values of one or more attributes, reflecting very small values for DEA efficiencies or probabilities of preference. Then, with denominators close to zero, the Malmquist ratios can vary widely. In the case of probabilistic composition, the occurrence of probabilities close to zero can be reduced by working with the probability of not being the best instead of the probability of being the best (or not being the worst instead of being the worst).

Thus, in the probabilistic approach, to avoid small values, a progressive index will employ the probability of not being the best by every criterion. This will be a negative score, decreasing as efficiency increases. It uses the formula for the optimistic score without subtraction from the minuend of 1.

In the conservative case, the corresponding index will follow the pessimistic composition, computing the probability of not being the worst by any criterion.

Assuming independence, the (decreasing) progressive index will be given by the product of the probabilities of not being the best and the (increasing) conservative index will be given by the product of the probabilities of not being the worst by each criterion. The use of the product is another factor of improving resistance against the influence of extreme values.

Finally, a compound probabilistic Malmquist index can be obtained by calculating the square root of the quotient between the conservative index (calculated by comparing efficiencies measured in terms of probabilities of not being the worst) and the progressive index (calculated by comparing probabilities of not being the best).

7.3 Malmquist Analysis of a Production Model

The present analysis considers the search for efficient operational costs for electricity distributors. The electric industry is characterized as a network industry. The joint and simultaneous action on three segments, generation, transmission and distribution of electricity, determines the final product available to the consumer.

In network industries, segments of transport and distribution are natural monopolies (Ramos-Real et al. 2009). To make the benefits of cost reduction due to economies of scale in the natural monopoly components of the network flow to price reduction for the consumer the interference of public power is employed.

To reduce appropriation by the monopolistic firms of extraordinary profits appears the figure of the regulator. Regulation must however balance the goal of reducing prices with encouraging economic efficiency and ensuring universality and quality of service. This can be achieved by simulating market conditions.

In Brazil the activity of electricity distribution is authorized by the Union conditioned upon signature of service contracts. Thus the electricity distributors serve its captive consumers in a Regulated Contracting Environment. The National Electric Energy Agency (ANEEL) regulates that market.

Among the regulatory powers of ANEEL is the Periodic Tariff Review, which is one of the mechanisms for setting the value of energy paid by electricity consumers. A review is held on average every 4 years. The Periodic Tariff Review repositions the value of the portion of the price that covers the cost of the distribution activity manageable by the distribution company. This mechanism should grant efficient levels of operational costs and adequate remuneration of the investments of the distributors. If formulated in a comparative basis, it may be able to replicate the advantages of a market environment.

According to ANEEL (2010), the variation of prices in a competitive market has three well-defined sources: technological change, economies of scale and scope and inputs prices. The first two of these factors impact prices by altering productivity. Thus, under the assumption of constant input prices, productivity increases must generate lower tariffs to be passed on to consumers. The crucial matter for the regulator is then establishing new standards of productivity.

Along the tariff cycles ANEEL has been refining the methodology used to determine the efficient operational costs. In the first tariff review cycle, cost was determined based on a reference company, a concept applied again in the second cycle, but with the introduction of an analysis of global consistency. The reference company is built as a mathematical representation of the distribution activity by defining the activities and processes of the distributors and the average cost of each activity performed.

The third cycle introduced the use of benchmark models. To take into account the evolution of efficiencies along time, in the benchmark models is common the use of indices such as Malmquist index. Jasmab and e Pollit (2001), reporting such increasing importance of benchmarking methods in the context of incentive regulation of distribution and transmission of electricity and analyzing OECD and some other countries, observe a preference for non-parametric methods.

ANEEL (2011) proposed to estimate the evolution of productivity using, as inputs, operational cost and, as outputs, size of the market served, in MWh/year offered, number of consumers in the region attended, and extent of the network, in Km.

Operational costs have considerable variation with a magnitude of millions of dollars per year for the larger Brazilian companies. The extension of the operated network reaches hundreds of thousands of miles and is a proxy for the costs of maintaining services. The number of consumers exceeds the total of a million in many cases, capturing the business costs incurred by the distributors. Finally, annual MWh captures the size of the distributor on issues where the voltage level affects the costs.

In ANEEL (2011) was used a decreasing returns to scale and input orientation DEA model. It is shown in the work of Andrade and Sant'Anna (2013), resumed here, how to replace DEA by the probabilistic composition, using the probabilistic Malmquist index to assess each company's effort on improving through time, taking into account its specific operating conditions.

The same variables in ANEEL (2011) study are employed here but the probabilities of maximization and minimization are calculated for the vectors of quotients of each output by the unique input. Thus, the variables employed are the v_{aj} below, v_{aj} denoting the ratio between the j-th output O_{aj} of the a-th company and its operational cost I_{aj}. It is assumed that, to measure efficiency, the v_{aj} are modes of independent random variables with triangular distribution of constant range.

$$v_{a1} = O_{a1}/I_{a1}, v_{a2} = O_{a2}/I_{a2}, v_{a3} = O_{a3}/I_{a3}.$$

Table 7.1 shows the values of the three ratios for each of the 61 distributors in the Brazilian electric system in the years 2008 and 2009. Table 7.2 presents first the 2009 efficiency scores for each company by the progressive and optimistic approach, followed by the scores for the pessimistic and conservative approach. Then comes the Malmquist index based on the frontier of excellence (the lower the score, the higher the evolution in efficiency) followed by the Malmquist index based on the lower frontier and finally the compound index.

Initially the probabilistic composition used is progressive and optimistic, which is conceptually similar to the benevolent approach of traditional DEA. The measure used for this approach is a measure of inefficiency, the probability of not being the

Table 7.1 Entrance data for comparative evaluation of electricity distribution firms

Firm	2008 Network	2008 Consumers	2008 Market	2009 Network	2009 Consumers	2009 Market
ADESA	0.0697	3.3213	7.7000	0.0694	3.0625	7.0501
AES SUL	0.3772	5.6967	14.1911	0.3694	5.5817	13.5724
AMPLA	0.1130	5.2915	7.7431	0.1169	5.4190	8.1977
BANDEIRANTE	0.0912	4.8205	15.0783	0.0959	5.1686	14.5826
BOA VISTA	0.0411	1.1557	3.6435	0.0427	1.2222	4.0580
CAIUÁ	0.1622	3.9706	9.5567	0.1857	4.5856	11. 3147
CEAL	0.1686	4.4931	4.9501	0.1451	3.7608	4.1373
CEB	0.0788	3.8795	9.7812	0.0769	3.7244	9.5560
CEEE	0.2055	4.3454	8.3704	0.1863	3.7257	7.2078
CELESC	0.1820	2.9631	8.4396	0.2008	3.1009	8.9545
CELG	0.3009	3.2537	5.3578	0.2885	3.2007	5.2983
CELPA	0.2426	4.6006	6.9274	0.2144	3.8586	5.4164
CELPE	0.3474	8.2711	10.4183	0.3434	8.5391	10.7434
CELTINS	0.4510	3.0114	4.9350	0.5054	3.3300	5.2151
CEMAR	0.4482	8.6068	9.4058	0.4657	8.7418	9.4780
CEMAT	0.3641	3.6479	8.0058	0.3707	3.4672	7.5837
CEMIG	0.2680	3.9361	8.0308	0.2736	4.0613	7.6902
CEPISA	0.2620	4.7348	4.7251	0.2350	4.3431	4.3234
CERON	0.0300	2.8660	5.7144	0.2217	2.7516	5.8000
CFLO	0.1281	3.5583	8.9636	0.1342	3.7480	9.2827
CHESP	0.3295	3.3022	5.4897	0.3251	3.0806	5.2834
CJE	0.0976	3.3652	21.8118	0.1071	3.7218	21.9248
CLFM	0.2126	5.1881	13.3765	0.2426	5.8507	14.7757
CLFSC	0.2298	4.4782	9.6074	0.2926	5.7111	12.5027
CNEE	0.1171	3.7123	9.3567	0.1224	3.8679	9.7234
COCEL	0.1866	3.3913	9.7659	0.1898	3.3881	9.5943
COELBA	0.4818	10.5197	11.9412	0.4926	10.5904	12.2414
COELCE	0.3467	8.0908	8.8056	0.3805	8.6816	9.6473
COOPERALIA	0.1312	3.1812	7.3645	0.1215	2.9416	6.8429
COPEL	0.2512	4.0090	9.2544	0.2207	3.5610	8.2155
COSERN	0.3513	8.4784	12.2528	0.3518	8.7437	12.2163
CPEE	0.2054	3.9831	11.9606	0.2449	4.7358	13.5520
CPFL PAULIS	0.1830	7.0478	17.9329	0.1807	7.0438	17.6908
CSPE	0.2829	5.2156	12.2940	0.3577	6.5977	15.1041

(continued)

Table 7.1 (continued)

Firm	2008 Network	2008 Consumers	2008 Market	2009 Network	2009 Consumers	2009 Market
DEMEI	0.0583	3.7427	7.8974	0.0564	3.7179	8.0501
DMEPC	0.0651	3.1120	9.4744	0.0575	2.8019	8.5529
EBO	0.2052	6.5322	10.3428	0.1865	6.0426	9.2770
EDEVP	0.2015	4.2458	8.9604	0.1950	4.1366	9.1076
EEB	0.1691	3.4073	8.4280	0.1988	3.9827	9.3714
EFLJC	0.0386	1.8838	4.3331	0.0337	1.7263	3.9247
EFLUL	0.0527	1.1271	10.2583	0.0576	1.2127	10.1872
ELEKTRO	0.2573	5.0273	11.2661	0.2584	5.1222	11.2633
ELETROACRE	0.2312	4.4095	8.3245	0.1796	3.1314	5.9301
ELETROCAR	0.2263	3.1070	7.7140	0.2240	3.1009	7.3251
ELETROPAULO	0.0430	5.5823	14.5719	0.0362	4.7936	12.3064
ELFSM	0.3639	4.2416	10.6615	0.3714	4.2840	10.8915
EMG	0.3571	5.2784	7.0163	0.3541	5.1888	6.9584
ENERSUL	0.3470	3.6609	6.8573	0.3680	3.8764	7.5527
ENF	0.0863	4.0119	7.8070	0.0897	4.1887	8.1142
EPB	0.3800	6.3672	6.7505	0.3852	6.2435	6.6706
ESCELSA	0.2423	5.0442	13.5212	0.2359	4.9100	11.3838
ESSE	0.2133	5.3986	8.4068	0.2408	5.7568	9.1493
FORCEL	0.0300	2.0570	5.8249	0.0300	2.0573	6.3503
HIDROPAN	0.0851	2.7421	8.1577	0.0857	2.8146	7.7477
IENERGIA	0.1385	1.8143	6.0646	0.1694	2.1160	7.2698
LIGHT	0.0953	5.8225	13.6933	0.1042	6.5329	14.8848
MUX-ENERGIA	0.1090	5.5205	15.2488	0.1087	5.4569	15.8080
PIRATININGA	0.1118	6.8397	22.8694	0.1136	6.9845	22.1597
RGE	0.4481	6.3653	16.2286	0.4561	6.5792	16.4037
SULGIPE	0.3241	5.8333	4.7737	0.2802	4.7243	3.8948
UHENPAL	0.2716	2.2409	5.2213	0.2651	2.1906	5.2537

Table 7.2 Static and dynamic indices for the Brazilian electric sector

Firm	2009	2009	2008/2009	2008/2009	2008/2009
	Optim. progr.	Pessim. conser.	Progr.	Conser.	Compound
ADESA	0.0250	0.9335	1.0008	0.9953	0.9973
AES SUL	0.0520	0.9838	1.0036	0.9993	0.9979
AMPLA	0.0310	0.9663	0.9992	1.0036	1.0022
BANDEIRANTE	0.0390	0.9673	1.0003	1.0039	1.0018
BOA VISTA	0.0220	0.7643	0.9998	1.0171	1.0086
CAIUÁ	0.0340	0.9755	0.9962	1.0062	1.0050
CEAL	0.0260	0.9042	1.0022	0.9648	0.9811
CEB	0.0280	0.9507	1.0005	0.9989	0.9992
CEEE	0.0290	0.9647	1.0026	0.9927	0.9950
CELESC	0.0300	0.9654	0.9989	1.0036	1.0024
CELG	0.0310	0.9467	1.0008	0.9983	0.9987
CELPA	0.0290	0.9507	1.0034	0.9816	0.9891
CELPE	0.0650	0.9846	0.9964	1.0005	1.0021
CELTINS	0.1570	0.9484	0.9214	1.0104	1.0471
CEMAR	0.1180	0.9845	0.9688	1.0003	1.0161
CEMAT	0.0410	0.9691	0.9988	0.9977	0.9995
CEMIG	0.0330	0.9708	1.0001	0.9995	0.9997
CEPISA	0.0300	0.9198	1.0020	0.9824	0.9902
CERON	0.0280	0.9477	0.9960	1.0914	1.0468
CFLO	0.0290	0.9647	0.9994	1.0017	1.0011
CHESP	0.0340	0.9459	1.0005	0.9943	0.9969
CJE	0.1590	0.9666	0.9939	1.0046	1.0054
CLFM	0.0450	0.9821	0.9939	1.0030	1.0045
CLFSC	0.0420	0.9821	0.9904	1.0072	1.0085
CNEE	0.0290	0.9643	0.9994	1.0014	1.0010
COCEL	0.0300	0.9682	1.0002	0.9997	0.9998
COELBA	0.2570	0.9882	0.9875	1.0004	1.0065
COELCE	0.0720	0.9839	0.9840	1.0022	1.0092
COOPERALIA	0.0260	0.9502	1.0008	0.9937	0.9965
COPEL	0.0300	0.9682	1.0028	0.9947	0.9959
COSERN	0.0720	0.9859	0.9945	1.0001	1.0028
CPEE	0.0390	0.9793	0.9949	1.0047	1.0049

(continued)

Table 7.2 (continued)

Firm	2009	2009	2008/2009	2008/2009	2008/2009
	Optim. progr.	Pessim. conser.	Progr.	Conser.	Compound
CPFL PAULIS	0.0610	0.9818	1.0018	0.9999	0.9991
CSPE	0.0570	0.9856	0.9817	1.0048	1.0117
DEMEI	0.0270	0.9321	0.9999	1.0003	1.0002
DMEPC	0.0260	0.9258	1.0012	0.9857	0.9922
EBO	0.0360	0.9762	1.0037	0.9970	0.9966
EDEVP	0.0310	0.9720	1.0001	0.9998	0.9999
EEB	0.0310	0.9717	0.9975	1.0067	1.0046
EFLJC	0.0220	0.7805	1.0004	0.9513	0.9751
EFLUL	0.0260	0.8479	0.9998	1.0176	1.0089
ELEKTRO	0.0380	0.9794	0.9998	1.0002	1.0002
ELETROACRE	0.0270	0.9515	1.0053	0.9785	0.9866
ELETROCAR	0.0290	0.9622	1.0006	0.9982	0.9988
ELETROPAULO	0.0330	0.9116	1.0064	0.9970	0.9953
ELFSM	0.0450	0.9787	0.9979	1.0005	1.0013
EMG	0.0410	0.9727	1.0017	0.9994	0.9989
ENERSUL	0.0410	0.9713	0.9959	1.0044	1.0043
ENF	0.0280	0.9548	0.9995	1.0016	1.0011
EPB	0.0500	0.9734	0.9981	0.9994	1.0006
ESCELSA	0.0360	0.9785	1.0035	0.9982	0.9973
ESE	0.0360	0.9777	0.9973	1.0033	1.0030
FORCEL	0.023	0.853	1.000	1.005	1.002
HIDROPAN	0.026	0.944	1.000	1.000	0.999
IENERGIA	0.026	0.940	0.998	1.026	1.013
LIGHT	0.044	0.969	0.994	1.001	1.003
MUX-ENERGIA	0.043	0.971	0.998	1.000	1.001
PIRATININGA	0.180	0.974	1.007	1.000	0.996
RGE	0.099	0.987	0.985	1.000	1.007
SULGIPE	0.033	0.900	1.006	0.959	0.976
UHENPAL	0.029	0.928	1.071	0.999	0.966

best by every criterion given by the product of the probabilities of not being the best according to each criterion. In this case, the value of the Malmquist index greater than unity represents an increase of inefficiency compared to the previous period. Instead, values below 1 indicate increased efficiency.

In Table 7.2 it can be seen that many companies present similar efficiency values, both in the static and the dynamic evaluations. In the 2009 analysis, only twelve companies present a score above the threshold of 0.05 in the progressive approach and only eighteen present a score lower than 0.95 in the conservative approach. From 2008 to 2009, according to neither calculation of the Malmquist index there were more than twelve companies outside the interval from 0.975 to 1.025.

The difference between the two approaches is enlightened by the scores of CELTINS, which, serving a small market spread on a large region, derives from its ample network the best score when the optimistic progressive algorithm is applied and bad scores when the pessimistic conservative approach looks for good performance with respect to all the criteria.

As an example of how the calculation is developed, Table 7.3 presents values of COELBA evaluations. This distributor showed remarkable growth along the first decade of this century without raising operational costs. While the operational cost fell 3.6 % from 2003 to 2009, its network expanded 67 %, the number of customers increased 38 % and the market third variable grew 42 %.

In the first lines of Table 7.3, the entries 0.4818, 10.5197 and 11.9413 of COELBA in Table 7.1 can be obtained dividing the absolute values of 201862,

Table 7.3 Computations in the determination of Malmquist indices for COELBA

	Operational Cost (R$1,000.00)	Network (km)	Consumers	Market (MWh)
2008 absolute	418,981	201,862	4,407,561	5,003,158
2008 relative		0.4818	105.197	119.413
2009 absolute	436,436	215,001	4,622,046	5,342,574
2009 relative		0.4926	105.904	122.414
Probabilities relative to 2008 frontier				
2008 max		0.1373	0.1629	0.0143
2008 min		0.0028	0.0037	0.0068
2009 max		0.1423	0.1635	0.0147
2009 min		0.0027	0.0037	0.0065
Probabilities relative to 2009 frontier				
2009 max		0.1011	0.1617	0.0141
2009 min		0.0030	0.0035	0.0053
2008 max		0.0857	0.1610	0.0137
2008 min		0.0031	0.0036	0.0055

4407561 and 5003158 of network, consumers and energy delivered in 2008 by the operational cost of 418,981. Analogously for 2009.

The progressive efficiency score for 2009 of 0.2570 in Table 7.2 is the complement to one of the product of the complements to one of the probabilities of maximizing the outputs in 2009, given by 0.1011, 0.1617 and 0.0141. In fact, 0.02570 = 1 − (1 − 0.1011)*(1 − 0.1617)*(1 − 0.0.141).

On the other hand, the conservative efficiency score of 0.9882 is the result of the product (1 − 0.0030)*(1 − 0.0035)*(1 − 0.0053).

Analogously, for 2008, the progressive and conservative scores are derived, respectively, from the probabilities of maximization, of 0.1373, 0.1629 and 0.0143 and of minimization, of 0.0028, 0.0037 and 0.0068.

The Malmquist progressive score is the square root of the product of two ratios.

The first ratio, relative to the frontier of 2008, is between the product (1 − 0.1423)*(1 − 0.1635)*(1 − 0.0147) derived from the probabilities of the entries 2009 reaching the frontier of 2008 and the product of the probabilities of not maximizing in 2008, (1 − 0.1373)*(1 − 0.1629)*(1 − 0.0143).

The second ratio, relative to the frontier of 2009, is between the product (1 − 0.1011)*(1 − 0.1617)*(1 − 0.0.141) and the product of the probabilities of the entries of 2008 not reaching the frontier of 2009, given by (1 − 0.0857)* (1 − 0.1610)*(1 − 0.0137).

Analogously, the Malmquist conservative scores are obtained applying the probabilities of minimization.

References

Andrade, G. N., & Sant'Anna, A. P. (2013). Composição probabilística e indice de Malmquist para avaliação de eficiência em distribuidoras de energia elétrica. *Anais do SBPO, 2013*, 937–948.

ANEEL. (2010). Metodologia de Cálculo dos Custos Operacionais, Nota Técnica n° 265/2010-SRE/ANEEL.

ANEEL. (2011). Metodologia de Cálculo dos Custos Operacionais, Nota Técnica n° 294/2011-SRE/ANEEL.

Caves, D. W., Christensen, L. R., & Diewert, W. E. (1982). The economic theory of index numbers and the measurement of input, output and productivity. *Econometrica, 50*(6), 1393–1414.

Entani, T., Maeda, Y., & e Tanaka, H. (2002). Dual models of interval DEA and its extensions to interval data. *European Journal of Operational Research, 136*, 32–45.

Fare, R., Grosskopf, S., & e Lovell, C. A. K. (1994). *Production frontiers*. Cambridge: Cambridge University Press.

Jasmab, T., & e Pollit, M. (2001). Benchmarking and regulation: international electricity experience. *Utilities Policy, 9*(3), 107–130.

Malmquist, S. (1953). Index numbers and indifference curves. *Trabajos de Estatistica, 4*(1), 209–242.

Ramos-Real, F. J., Tovar, B., Iootty, M., de Almeida, E. F., & Pinto, H. Q, Jr. (2009). The evolution and main determinants of productivity in Brazilian electricity distribution 1998–2005: An empirical analysis. *Energy Economics, 31*, 298–305.

Yamada, Y., Matui, T., & e Sugiyama, M. (1994). New analysis of efficiency based on DEA. *Journal of the Operations Research Society of Japan, 37*(2), 158–167.

Chapter 8
Probabilities in the Problem of Classification

Abstract Evaluation of alternatives may be derived from probabilistic comparisons made with different sets of profiles which represent ordered classes. This may be used to solve the problem of, instead of choosing the best or worst alternative, identifying alternatives that correspond to each of a set of different levels of performance.

Keywords Classification · Interval classification · Profile · PseudoF · IDH

8.1 Modeling the Problem of Classification

It was demonstrated in the preceding chapters that, while the computation of the probabilistic scores of preference given by probabilities of being the best is easier if the comparisons are made to a sample of fixed previously chosen representative alternatives instead of to the entire population of alternatives, the efficiency of the evaluation will increase with the number of comparisons made. For these reasons, it is important to balance the goal of reducing the number of comparisons to simplify computations with that of increasing it to improve reliability.

In the problem of classification, addressed in this chapter, a large number of comparisons is made, but the decisions are taken successively based on comparisons to small sets of possible alternatives previously ordered.

As in the problem of choice of the best or the worst, in the problem of classification the probabilistic approach is based on treating the initial numerical evaluations as location parameters of probability distributions.

To formulate the problem of classification the following terms are employed.

- $G = \{g_1, \ldots, g_m\}$, a set of m criteria.
- $A = (a_1, \ldots, a_m)$ a vector of R^m which stores the evaluations according to the m criteria of the alternative to be classified; the highest the value of the coordinate a_j, the better the alternative according to the criterion g_j.
- $C = \{C_1, \ldots, C_r\}$ a set of r classes, ordered from the worst to the best, so that the alternative is better if classified in a class of higher index.

A. Parracho Sant'Anna, *Probabilistic Composition of Preferences, Theory and Applications*, Decision Engineering,
DOI 10.1007/978-3-319-11277-0_8

To identify the class C_i, for each i from 1 to r are employed n(i) representative profiles, each of them formed by a vector of evaluations by the m criteria.

The *h*th representative profile of the *i*th class is given by a vector of evaluations by the m criteria $(c_{ih1}, \ldots, c_{ihm})$.

The representative profiles are built in such a way that if $i_1 < i_2$, then, for every j, $c_{i1hj} \leq c_{i2hj}$ and, for at least one j_o, $c_{i1hjo} \leq c_{i2hjo}$.

So, according to no criterion a profile of a higher class may present an evaluation lower than that of a profile of a lower class by that same criterion and by at least one criterion the profile of the higher class presents evaluation higher than that of the profile of the lower class.

To allow for the probabilistic comparison, the evaluations a_j of the alternative being classified are replaced by probability distributions centered at a_j. To classify the alternative what is going to be compared are the probabilities of the alternative presenting evaluations according to such distributions above or below the profiles of each class. The c_{ihj} may be also replaced by random variables X_{ihj}.

This characterization of the problem of classification follows that of Roy (1968) and Yu (1992). The representative profiles approach follows Almeida Dias et al. (2010, 2012). More details about such characterization are available in Sant'Anna et al. (2012).

8.2 Computation of Probabilities of Preference

Following the principles of classical statistical modeling, it is reasonable to assume here, as in the preceding chapters, not only a normal form, but also identical distributions and independence between the disturbances causing the imprecision in the evaluations according to any criterion. Alternatively, instead of normal, triangular distributions may be a simpler starting point, like in the Theory of Fuzzy Sets (Zadeh 1965). In advancing the analysis, the kind of evaluation involved may also suggest a different distribution.

Since the mean of the normal distribution or the mode of the triangular distribution are determined by the observed values, to complete modeling the distributions in these two cases it is enough to determine the values, respectively, of the variance and of the extreme points. The available data are used to estimate these parameters or other parameters that determine other distributions. If other information on the parameters is available in advance, it can be used and will simplify computations.

Assuming normality, for each j from 1 to m, the coordinates a_j of the alternative evaluated, for i from 1 to r, are considered as means of independent normal distributions with the same variance.

If the variance is small, is small the probability that an alternative with evaluation a_j according to the *j*th criterion may belong to the classes with profiles presenting for that criterion values more distant to a_j. So, the variance must be large enough to make that probability not so small that the information given by the other criteria may become irrelevant, giving veto power to a single criterion.

This principle can be satisfied, in practice, if we estimate the variance of the evaluations according to each criterion by the variance of the set of values by this criterion in the various profiles offered as representative of the classes. This may lead to overestimation, because the variance in such a sample combines the inner variance in each class with the variance between classes and only the first of these components must be attributed to the uncertainty inherent in each register. In fact, each alternative should belong to a proper class and differences between classes are not of a random nature. But a possible overestimation is welcome to avoid the above mentioned risk of observations being deemed as too far away from some of the reference profiles suggested.

Once replaced the exact measurements a_j and C_{ihj} by distributions of random variables X_k and Y_{ihj} centered on these measurements, probabilities of outranking can be calculated. Let us denote by A_{ij}^{+} and A_{ik}^{-} the probabilities of the alternative A presenting a value for X_j respectively above and below those associated to the *j*th criterion in the *i*th class profiles. By independence between disturbances affecting the evaluations by a given criterion of different alternatives and different profiles,

$$A_{ij}^{+} = \prod_{h} P\left[X_j > Y_{ihj}\right]$$

and

$$A_{ij} = \prod_{h} P\left[X_j < Y_{ihj}\right].$$

To make sure that the probability of an alternative being above the representative profiles decreases as increases the order of the class as well as, conversely, the probability of being below increases, it is necessary that the number of profiles be the same in all classes, i.e., that, for all i, n(i) = k, for a constant k.

If the decision maker, when modeling the classes, offers different numbers of profiles, additional profiles may be assigned to the classes with number of profiles different of the maximum. To form these complementing profiles, median or mean values may be employed.

8.3 Composition Rules

The Choquet integral with respect to a capacity that reflects an assessment of the importance of the criteria can be used to compose global evaluations of the alternative being above and being below each class. Assuming linearity, this can be reduced to a weighted average.

If there is no way to differentiate the relative importance of the criteria, simpler compositions, like those by joint probabilities, may be employed In that case the extremes of independence and of maximal dependence result in simple formulae.

The assumption of maximal dependence leads to simpler calculations than that of independence. The foundation for this assumption may be strengthened in the problem of classification by the fact that the criteria are made to agree with the ordering of the classes. However, it should be taken into account that correlation between the vectors of evaluations of the different criteria does not imply dependence between the disturbances that assign the random character to the measurements.

On another direction, the agreement of the evolution of the values of the criteria in the profiles along the classes, implying that the criteria confirm each other, is a reason to take a pessimistic approach. In that case, the combination of the criteria by joint probability will use, for the probability of the alternative being above the *i*th class the formulae:

$$A_i^+ = \min_j A_{ij}^+, \text{ for the hypothesis of maximal dependence}$$

or

$$A_i^+ = \pi_j A_{ij}^+, \text{ for the hypothesis of independence.}$$

And for the probability of the alternative being below the *i*th class

$$A_i^- = \min_j A_{ij}^-$$

or

$$A_i^- = \pi_j A_{ij}^-.$$

The composition formulae for the Choquet integral will be:

$$A_i^+ = \sum_{j=1}^{m} (A_{i\tau i+(j)+} - A_{i\tau i+(j-1)+})\mu(\{\tau i^+(j), \ldots, \tau i^+(m)\})$$

and

$$A_i^- = \sum_{j=1}^{m} (A_{i\tau i-(j)-} - A_{i\tau i-(j-1)-})\mu(\{\tau i^-(j), \ldots, \tau i^-(m)\})$$

where μ is the capacity,

$A^+_{i\tau i+(0)} = A^-_{i\tau i-(0)} = 0$, for all i,

and τi^+ and τi^- are the permutations of the set of indices {1, …, m} such that

$$A^+_{i\tau i+(1)} \le A^+_{i\tau i+(2)} \le \cdots \le A^+_{i\tau i+(n-1)} \le A^+_{i\tau i+(n)}$$

and

$$A^-_{i\tau i-(1)} \le A^-_{i\tau i-(2)} \le \cdots \le A^-_{i\tau i-(n-1)} \le A^-_{i\tau i-(n)}.$$

For weighted average:

$$A_i^+ = S_j w_j A_{ij}^+$$

and

$$A_i^- = S_j w_j A_{ij}^-,$$

for w_j denoting the weight assigned to the *j*th criterion.

8.4 The Algorithm of Classification

The classification procedure is based on the comparison of the differences $A_i^+ - A_i^-$.

The classification rule is simple: alternative A belongs to the class i with the smallest absolute value for $A_i^+ - A_i^-$.

Since there is the possibility of ties, the alternative may be classified in two adjacent classes. With the distributions and profiles constructed as above described, no more than two classes can present the same absolute value for the difference $A_i^+ - A_i^-$.

If the criteria obey the principle of assigning values increasing with the preference for the alternative, these differences constitute a non-increasing sequence. Then, an efficient algorithm for applying this rule can be developed in two steps. In the first step, it identifies the smallest value of i for which the difference $A_i^+ - A_i^-$ is negative or null. If this difference is always positive, the alternative must belong to the highest class. If it is never positive, this value is 1 and the alternative is placed in the lowest class. If the difference is zero for the class with the smallest absolute value, then the alternative is placed in that class.

If none of these cases happen, a second step consists in comparing the absolute values of the differences $A_i^+ - A_i^-$ for the class with the first negative difference and for that one preceding it. If these absolute values are the same, the alternative is classified in these two classes. If they are different, it is placed in that one of these two classes with the smallest absolute value for the difference.

Theoretically, due to continuity, possibility of ties can be neglected. But, if the number of criteria and the number of values for each criterion is not too small, ties may be bring to happen in practice if approximations make small the numerical differences, specially when the minimum is used to combine evaluations, in the joint probabilities composition under the assumption of dependence. The possibility of allocation in a pair of classes or even in a larger interval of acceptable values of i is a natural consequence of the imprecision of the subjective process of determining preferences.

A procedure to automatically determine the extremes of such an interval may be formalized by following a descending procedure that stops at the highest possible point, to determine the highest extreme, and an ascending procedure that stops at the lowest possible point, to determine the lowest one. The ascending procedure will be named in what follows the benevolent procedure, and the descending one the hostile procedure.

Denoting by U(A) the class in the upper extreme of the set where alternative A is classified and by L(A) the class in the lower extreme, let us start by the hostile procedure.

The first step consists in assigning a provisional lower classification

$$LP(A) = \min\{i : A_i^+ - A_i^- \leq 0\}.$$

To determine this value, start with LP(A) = r; if $A_r^+ - A_r^- \leq 0$, try r − 1 and keep reducing the value of i until this is no longer true.

If $A_r^+ - A_r^- > 0$, then$\{i: A_i^+ - A_i^- \leq 0\}$ is empty and the alternative belongs to class C_r, the highest class,

$$L(A) = C_r.$$

The classification is then punctual and the benevolent procedure needs no longer be applied.

If LP(A) = 1, then the lower end of the interval is given by class C_1,

$$L(A) = C_1.$$

For $\{i: A_i^+ - A_i^- \leq 0\}$ not empty and LP(A) > 1,

if $A^-_{LP(A)} - A^+_{LP(A)} < A^+_{LP(A)-1} - A^-_{LP(A)-1}$, then

$$L(A) = C_{LP(A)}.$$

Otherwise, if LP(A) − 1 = 1, then the process finishes with

$$L(A) = C_1,$$

and, if LP(A) − 1 > 1, compare $A^+_{LP(A)-1} - A^-_{LP(A)-1}$ with $A^+_{LP(A)-2} - A^-_{LP(A)-2}$.

If $A^+_{LP(A)-1} - A^-_{LP(A)-1} < A^+_{LP(A)-2} - A^-_{LP(A)-2}$, then

$$L(A) = C_{LP(A)-1}.$$

If $A^+_{LP(A)-1} - A^-_{LP(A)-1} = A^+_{LP(A)-2} - A^-_{LP(A)-2}$, then substitute LP(A) − 2 for LP(A) − 1 as the provisional lower extreme and repeat the preceding step.

The benevolent procedure follows the same route, starting by a provisional upper bound

$$UP(A) = \max\{i: A_i^- - A_i^+ \leq 0\}.$$

If $\{i: A_i^- - A_i^+ \leq 0\}$ is empty, then the alternative belongs to the lowest class,

$$U(A) = L(A) = C_1.$$

If $UP(A) = r$, then the upper end of the interval is given by class C_r,

$$U(A) = C_r.$$

For $\{i: A_i^- A_i^+ \leq 0\}$ not empty and $UP(A) < r$,
if $A^+_{UP(A)} - A^-_{UP(A)} < A^+_{UP(A)+1} - A^-_{UP(A)+1}$, then

$$U(A) = C_{UP(A)}.$$

Otherwise,
if $UP(A) + 1 = r$, then

$$U(A) = C_r,$$

and,
if $UP(A) + 1 < r$,
compare $A^+_{UP(A)-1} - A^-_{UP(A)-1}$ with $A^+_{UP(A)-2} - A^-_{UP(A)+2}$.
If $A_{UP(A)+1}{}^- A^+_{UP(A)+1} < A^-_{UP(A)+2} - A^+_{UP(A)+2}$, then

$$U(A) = C_{UP(A)+1}.$$

If $A^+_{UP(A)+1} - A^-_{UP(A)+1} = A^+_{UP(A)+2} - A^-_{UP(A)+2}$, then substitute $UP(A) + 2$ for $UP(A) + 1$ as the provisional lower extreme and repeat the preceding step.

It is easy to see that the benevolent procedure so defined will stop at a higher or equal classification than the hostile one. In fact, the descending procedure stops at the highest i minimizing $A_i^+ - A_i^-$ and the ascending procedure stops at the lowest i minimizing that difference.

As already explained, unless rough approximations are employed, the minimum will be unique and the benevolent and the hostile classification will coincide. As this uniqueness of the classification may be unsatisfactory, because uncertainty is a feature of preference measurements, one way to provide more information on the subjacent uncertainty consists in enlarging the interval by making more benevolent the benevolent procedure and more hostile the hostile procedure.

More benevolent and more hostile classifications will be based respectively on more stringent rules for classifying below and above the profiles of each class. Such rules may be determined by applying fixed rates of reduction respectively to the probability of being below the representative profiles of each class in the descending procedure and to the probability of being above in the ascending procedure.

A benevolent classification with cutting points determined by the reduction percent c will place alternative A in the class C_i for that i minimizing the absolute value of the difference $A_i^+ - (1 - c)A_i^-$.

Analogously, a hostile classification for the same reduction percent will place alternative A in that class C_i for the i determined by the hostile procedure modified by replacing the difference $A_i^+ - A_i^-$ with $(1 - c)Ai^+ - A_i^-$.

Another way to generate more benevolent and more hostile extremes without the need to determine a rate of reduction may be applied if the combination of the criteria is by joint probabilities. In that case, to obtain the benevolent classification, the joint probability of being above all the profiles of the class may be replaced by the probability of being above at least one of them. On the other end, in the computation of the hostile extreme, the joint probability of being below all the profiles of the class will be replaced by the probability of being below at least one of them.

That means, assuming maximal dependence, to get the benevolent extreme in the computation of A_i^+, substituting $\max_j A_{ij}^+$ for $\min_j A_{ij}^+$ and, to get the hostile extreme, substituting $\max_j A_{ij}^-$ for $\min_j A_{ij}^-$. In the case of independence, instead of $\max_j A_{ij}^+$ and $\max_j A_{ij}^-$, respectively, $1 - \pi_j(1 - A_{ij}^+)$ and $1 - \pi_j(1 - A_{ij}^-)$ enter here.

To conclude this section a practical hint on the generation of profiles may be useful. In practice, we have to classify not a unique alternative, but a set of different alternatives, as in the ranking problem. This may be used in the process of generating profiles. For instance, central profiles for r classes would be formed with coordinates given by the quantiles of order 1/2r, 3/2r, …, (2r − 1)/2r, for every criterion.

8.5 Classification of Car Models

The classification setup may be applied to the data of 20 car models studied in the previous chapters. Suppose that there are five classes where they might be classified each identified by four profiles, given in Table 8.1.

The probabilities of each alternative being above and below all the four profiles for each of the 5 classes are presented in Table 8.2, assuming triangular distributions with extremes 0 and 1 and composing the criteria by a weighted average with the weights derived in Chap. 2 from pairwise comparison: 0.03 for beauty and for power, 0.05 for price, 0.08 for comfort, 0.15 for gas consumption and 0.33 for reliability and for safety.

The last column of this table presents the classification obtained. Only Car3, Car16 and Car17 are classified in the highest class and no alternative is classified in the lowest.

If instead of triangular distributions, normal distributions with standard deviations of 0.4—an approximate value for the observed standard deviations in the samples of all the different criteria—are employed, the classification obtained is

Table 8.1 Representative profiles for 5 categories

	Beauty	Comfort	Consumption	Power	Price	Reliability	Safety
c1							
c11	0	0	0	0	0	0	0
c12	0	0	0	0	0	0.5	0
c13	0	0	0	0	0	0	0.5
c14	0	0	0	0	0	0.5	0.5
c2							
c21	0	0	0	0	0	1	0.5
c22	0.5	0	0	0	0	0.5	0.5
c23	0.5	0	0	0	0	1	0.5
c24	0.5	0	0	0	0	1	0.5
c3							
c31	0.5	0	0	0	0	1	1
c32	0.5	0.5	0	0	0	1	1
c33	1	0.5	0	0	0	1	1
c34	0.5	1	0	0	0	1	1
c4							
c41	1	1	0	0.5	0	1	1
c42	1	1	0	0	0.5	1	1
c43	1	1	0	0.5	0.5	1	1
c44	1	1	0	1	0.5	1	1
c5							
c51	1	1	0.5	1	0.5	1	1
c52	1	1	0.5	1	1	1	1
c53	1	1	1	1	0.5	1	1
c54	1	1	1	1	1	1	1

almost the same. Only Car11 and Car13 move, from the second and the fourth class, respectively, to the 3rd class.

For the classification combining the criteria by joint probabilities, assuming maximal dependence, instead of a unique class as in Table 8.2, interval classifications appear. Table 8.3 presents three interval classifications derived from the probabilities of being above and below the profiles of each class calculated employing the triangular distribution.

The first is the basic classification comparing the estimates A_i^+ and A_i^- given, respectively, by the minimum of the probabilities of being above and below by each criterion. Then come the classifications resulting from application of rates of reduction of 0.5. And, finally, those resulting from comparison of the minima by one side to the maxima by the other.

Table 8.2 Probabilistic scores and final classification by weighted average

	Class 1		Class 2		Class 3		Class 4		Class 5		Final
	Above	Below	Above	Below	Above	Below	Above	Below	Above	Below	
Car1	0.41	0.12	0.30	0.22	0.24	0.23	0.21	0.31	0.15	0.33	3
Car2	0.41	0.12	0.36	0.23	0.24	0.29	0.21	0.31	0.15	0.33	3
Car3	0.52	0.05	0.41	0.11	0.28	0.10	0.25	0.19	0.20	0.22	5
Car4	0.47	0.07	0.36	0.13	0.23	0.12	0.20	0.20	0.17	0.27	4
Car5	0.34	0.15	0.23	0.25	0.17	0.26	0.14	0.34	0.11	0.40	2
Car6	0.34	0.15	0.29	0.26	0.17	0.32	0.14	0.34	0.11	0.40	2
Car7	0.48	0.07	0.37	0.13	0.24	0.12	0.21	0.20	0.18	0.26	4
Car8	0.35	0.15	0.24	0.25	0.18	0.26	0.15	0.33	0.12	0.39	2
Car9	0.46	0.08	0.35	0.14	0.22	0.13	0.19	0.21	0.16	0.28	4
Car10	0.50	0.06	0.39	0.12	0.27	0.12	0.25	0.21	0.19	0.24	4
Car11	0.38	0.14	0.33	0.25	0.22	0.31	0.20	0.34	0.14	0.37	2
Car12	0.46	0.08	0.35	0.14	0.23	0.14	0.21	0.22	0.17	0.28	4
Car13	0.44	0.09	0.32	0.14	0.21	0.15	0.19	0.24	0.15	0.30	4
Car14	0.45	0.08	0.33	0.14	0.22	0.14	0.20	0.23	0.16	0.29	4
Car15	0.41	0.12	0.31	0.23	0.25	0.24	0.22	0.31	0.15	0.33	3
Car16	0.52	0.05	0.41	0.11	0.28	0.11	0.26	0.19	0.20	0.22	5
Car17	0.53	0.05	0.42	0.11	0.29	0.10	0.27	0.18	0.21	0.21	5
Car18	0.46	0.08	0.35	0.14	0.22	0.13	0.20	0.22	0.16	0.28	4
Car19	0.49	0.07	0.38	0.13	0.26	0.13	0.24	0.22	0.19	0.25	4
Car20	0.45	0.08	0.33	0.14	0.22	0.15	0.20	0.23	0.16	0.29	4

Table 8.3 Interval classifications by joint probabilities

	Balanced Interval		0.5 Reductions Interval		Max and Min Interval	
	Lower	Upper	Lower	Upper	Lower	Upper
Car1	3	4	3	4	1	5
Car2	3	4	2	4	1	5
Car3	4	4	4	5	2	5
Car4	4	4	4	5	2	5
Car5	3	3	3	3	1	5
Car6	3	3	2	3	1	5
Car7	4	4	4	4	2	5
Car8	3	3	3	4	1	5
Car9	4	4	3	4	2	4
Car10	4	4	4	4	2	5
Car11	3	4	2	4	1	5
Car12	4	4	3	4	2	5
Car13	3	3	3	4	2	5
Car14	4	4	3	4	2	5
Car15	3	4	3	4	1	5
Car16	4	4	3	4	1	5
Car17	4	4	3	4	1	5
Car18	3	3	3	4	1	5
Car19	4	4	3	4	1	5
Car20	3	3	3	4	1	5

It can be seen in Table 8.3, as the range of the intervals increases, that the classification of some cars can move to a better or a worse class. For instance, looking at the central columns, it can be detected a possibility of Car8, Car13, Car18 and Car20 to improve their classifications from C_3 to C_4 and a possibility of Car9, Car16, Car17 and Car19 to suffer a downgrade from C_4 to C_3.

8.6 Classification of Countries by HDI Criteria

An example of generation of profiles by the sample of alternatives to be classified may be given by revisiting the analysis of human development of countries on the basis of the four criteria: longevity, mean years of schooling, expected years at school and per capita income. Suppose that, instead of determining a distance to a global frontier, we prefer to classify the countries in a small number of classes. Fixing in five the number the classes, one profile for each class may be derived from the deciles of the sample of observed values.

By similar procedures, other two profiles may be derived for each class. In addition to the vectors of coordinates given by approximate values of the first, third, fifth, seventh and ninth deciles for each criterion separated, corresponding to the ranks 19th, 56th, 93rd, 131st and 168th in the set of 187 countries, another set of profiles is formed by the observed values of the criteria in five countries ranked in these five positions by the IDH (Austria, Romania, Iraq, Algeria and Côte d'Ivoire). A third set is formed with equally spaced values approximating the values observed in these two sets of profiles.

The vectors of values according to each criterion in the set of profiles so generated, shown in Table 8.4, have standard deviations of approximately 7.5, 3.3 and 10,000. The classifications of the 30 countries of Table 6.2, obtained by combining the criteria by joint probabilities, assuming maximal dependence and normal disturbances with these standard deviations, are shown in Table 8.5.

The first is the basic classification, obtained comparing the estimates A_i^+ and A_i^- for the probabilities given by the minimum of the respective probabilities of being above and below by each criterion. The ascending and descending procedures led then to the same classifications.

After that is the classification resulting from application of rates of reduction of 0.5. And finally the classification resulting from comparison of the minima of being by one side of the classes to the maxima of being by the other.

It can be seen in Table 8.5 that 20 of the 30 classifications are not subject to variation. By the application of the proportional reduction, there is doubt only about the classification of Equatorial Guinea, which may stay in the second class or be moved to the third. Finally, in the enlargement associated to employing

Table 8.4 Profiles of 5 classes for countries sorting

	Longevity	Mean schooling	Expected schooling	Income
9th decile	81	12	16	34,000
Austria	81	11	15	36,000
Class 5	76	12	16	18,000
7th decile	76	10	14	14,000
Romania	74	10	15	11,000
Class 4	74	10	14	14,000
5th decile	73	8	13	8,000
Algeria	74	8	14	7,000
Class 3	72	8	12	10,000
3rd decile	68	6	11	3,000
Iraq	70	6	10	4,000
Class 2	70	6	10	6,000
1st decile	55	3	9	1,000
Côte d'Ivoire	56	4	7	2,000
Class 1	68	4	8	2,000

probabilities of outranking the profiles in at least one criterion, large intervals are obtained for three more countries, Liechtenstein, Qatar and Kuwait.

All these results are explained by the construction of the sample. The four countries with large intervals are those with strongly divergent classifications by different criteria. The ten countries classified in the highest class without any doubt are exactly those ten countries with the highest HDI. On the other end, the eight last lines in the table, with coincident benevolent and hostile classification in the lowest class, correspond to the countries with the eight lowest HDI values.

The results of the application of more complex forms of composition are presented in Table 8.6. The first applies the additive composition with constant weights. In the second, is employed again a capacity built by assuming additivity and equal importance, but treating as mutually interchangeable the two educational indicators. In the third, this second capacity is modified, becoming interchangeable not only the two educational components but also the two indicators of longevity and income. Assuming equal importance for each of these two sets of two elements, numerically the capacity is formed by each set with one element of each pair of interchangeable components having a value of 1, while the unitary sets as well as the sets of two elements of the same interchangeable pair have the value of 1/2.

For these three capacities, the balanced classification is punctual. The benevolent and hostile classifications by reductions of 50 % are shown in Table 8.6 flanking the balanced classification. There is again high agreement between classifications, a divergence of more than one level never occurring between classifications in this table or between a classification of this table and a composition by joint probability in Table 8.5.

8.7 Evaluation of the Classification

To evaluate the homogeneity of the classes obtained after classifying a set of alternatives, the measure proposed by Calinski and Harabasz (1974) may be employed. It is based on the ratio between the sum of the variances within the clusters, of the vectors of evaluations around a center of the cluster, and the variance of these centers around the mean of all evaluations. Due to the conceptual similarity to the F distribution employed to evaluate normal regression, this measure is called PseudoF.

There are many different forms of determining the centers of the r classes and of combining distances on the m criteria.

Let us consider the centers of the classes, for i varying from 1 to r, given by the vectors $(\overline{c}_{i1}, \ldots, \overline{c}_{im})$ of averages of the evaluations of the alternatives in the classes and the general center given by the vector

$$\overline{c} = (\overline{c}_1, \ldots, \overline{c}_m)$$

of the averages of all the evaluations.

Table 8.5 Classification of countries by joint probabilities

	Balanced	0.5 Reductions Interval		Max and Min Interval	
		Lower	Upper	Lower	Upper
Norway	5	5	5	5	5
Australia	5	5	5	5	5
United States	5	5	5	5	5
Netherlands	5	5	5	5	5
Germany	5	5	5	5	5
New Zealand	5	5	5	5	5
Sweden	5	5	5	5	5
Switzerland	5	5	5	5	5
Japan	5	5	5	5	5
Hong Kong	5	5	5	4	5
Iceland	5	5	5	4	5
Israel	5	5	5	5	5
Singapore	5	5	5	4	5
France	5	5	5	4	5
Liechtenstein	5	5	5	3	5
Qatar	5	5	5	3	5
Kuwait	5	5	5	2	5
Equatorial Guinea	2	1	2	1	5
Papua New Guinea	1	1	1	1	2
Djibouti	1	1	1	1	1
Sudan	1	1	1	1	2
Sierra Leone	1	1	1	1	1
Central African R	1	1	1	1	1
Eritrea	1	1	1	1	1
Mali	1	1	1	1	1
Burkina Faso	1	1	1	1	1
Chad	1	1	1	1	1
Mozambique	1	1	1	1	1
D.R. Congo	1	1	1	1	1
Niger	1	1	1	1	1

Under this approach,

$$\overline{c}_{ij} = \sum_{a=1}^{n(i)} c_{\mathrm{iaj}}/\mathrm{n_i}$$

for $\mathrm{n_i}$ the number of alternatives classified in the *i*th class and $\mathrm{c_{iaj}}$ the evaluation of the *a*th of these alternatives by the *j*th criterion. Analogously,

Table 8.6 Classification of countries by Choquet integrals

	Four Dimensions			Three Dimensions			Two Dimensions		
	Hostile		Benevolent	Hostile		Benevolent	Hostile		Benevolent
Norway	5	5	5	5	5	5	5	5	5
Australia	5	5	5	5	5	5	5	5	5
United States	5	5	5	5	5	5	5	5	5
Netherlands	5	5	5	5	5	5	5	5	5
Germany	5	5	5	5	5	5	5	5	5
New Zealand	5	5	5	5	5	5	5	5	5
Sweden	5	5	5	5	5	5	5	5	5
Switzerland	5	5	5	5	5	5	5	5	5
Japan	5	5	5	5	5	5	5	5	5
Hong Kong	5	5	5	5	5	5	5	5	5
Iceland	5	5	5	5	5	5	5	5	5
Israel	5	5	5	5	5	5	5	5	5
Singapore	5	5	5	5	5	5	4	5	5
France	5	5	5	5	5	5	5	5	5
Liechtenstein	4	5	5	5	5	5	4	5	5
Qatar	3	5	5	4	5	5	3	4	5
Kuwait	3	4	5	4	5	5	3	4	5
Equatorial Guinea	1	1	2	1	2	3	1	1	2
Papua New Guinea	1	1	1	1	1	2	1	1	1
Djibouti	1	1	1	1	1	1	1	1	1
Sudan	1	1	1	1	1	1	1	1	1
Sierra Leone	1	1	1	1	1	1	1	1	1

(continued)

Table 8.6 (continued)

	Four Dimensions			Three Dimensions			Two Dimensions		
	Hostile		Benevolent	Hostile		Benevolent	Hostile		Benevolent
Central African R	1	1	1	1	1	1	1	1	1
Eritrea	1	1	1	1	1	1	1	1	1
Mali	1	1	1	1	1	1	1	1	1
Burkina Faso	1	1	1	1	1	1	1	1	1
Chad	1	1	1	1	1	1	1	1	1
Mozambique	1	1	1	1	1	1	1	1	1
D.R. Congo	1	1	1	1	1	1	1	1	1
Niger	1	1	1	1	1	1	1	1	1

$$\overline{c}_j = \sum_{i=1}^{r}\left(\sum_{a=1}^{\mathrm{n_i}} c_{\mathrm{iaj}}\right)/\sum_{i=1}^{r}\mathrm{n_i}$$

The PseudoF measure of quality of the classification is then given by the ratio between B/(r − 1) and W/(n − r) for B denoting the variance of the conditional expectation given the classification, W the expectation of the conditional variance, r the number of classes and n the number of alternatives classified.

Thus,

$$\mathrm{PseudoF} \;=\; (\mathrm{B}/(\mathrm{r}-1))/(\mathrm{W}/(\mathrm{n}-\mathrm{r}))$$

for

$$\mathrm{B} = \sum_{i=1}^{r}\mathrm{n_i}\sum_{j}\left(\overline{c}_{ij}-\overline{c}_j\right)^2,$$

$$\mathrm{W} = \sum_{i=1}^{r}\sum_{j}\sum_{a}\left(c_{\mathrm{iaj}}-\overline{c}_{ij}\right)^2,$$

and

$$\mathrm{n} = \sum_{i=1}^{r}\mathrm{n_i}.$$

Thus,

$$\mathrm{PseudoF} = \left(\sum_{i=1}^{r}\mathrm{n_i}\sum_{j}(\overline{c}_{ij}-\overline{c}_j)^2/\left(\sum_{i=1}^{r}\sum_{j}\sum_{a}(c_{\mathrm{iaj}}-\overline{c}_{ij})^2\right)/\left((\mathrm{r}-1)/\left(\sum_{i=1}^{r}\mathrm{n_i}\right)-\mathrm{r}\right)\right).$$

Given the decomposition of the variance,

$$\mathrm{PseudoF} = \left(\left(\sum_{i}\sum_{j}(c_{\mathrm{iaj}}-\overline{c}_j)^2-\sum_{i}\sum_{j}\sum_{a}(c_{\mathrm{iaj}}-\overline{c}_{ij})^2\right)/\sum_{i}\sum_{j}\sum_{a}(c_{\mathrm{iaj}}-\overline{c}_{ij})^2/\left((\mathrm{r}-1)/\sum_{i}\mathrm{n_i}-\mathrm{r}\right)\right)$$

This construction does not take into account the possibility of different scales of measurement of the different criteria. An alternative measure for the case of measures of importance for the criteria being available would compute the ratios separately for each criterion and weigh these ratios with weights that take into account such scales:

$$\text{WPseudoF} = \sum_{j=1}^{m} w_j (B_j/(r-1))/(Wj/(\sum_i \mathrm{n_i} - r))$$

for

w_j denoting the weight assigned to the *j*th criterion,
B_j denoting the variance of the conditional expectation for the *j*th criterion and
W_j denoting the expected value of the conditional variance for the *j*th criterion.
Thus,

$$\text{WPseudoF} = \sum_{j=1}^{m} w_j \left[\left(\sum_{i=1}^{r} \mathrm{n_i} \left(\overline{c}_{ij} - \overline{c}_j \right)^2 \right) / \left(\sum_{i=1}^{r} \sum_{a=1}^{\mathrm{n_i}} \left(c_{\mathrm{iaj}} - \overline{c}_{ij} \right)^2 \right) \right] / \left((\mathrm{r}-1) / \left(\sum_i \mathrm{n_i} - \mathrm{r} \right) \right).$$

Table 8.7 presents, for the classification in Table 8.3, the means and sums of squares for the four classes and the seven criteria—employed in the computation of W. Presents also the general means and the sums of squares relatively to these global means after replacing the original values by the means of the classes—employed in the computation of B.

Table 8.8 presents the values of PseudoF and WPseudoF derived from the values of Table 8.7 and for the case of Car11 and Car13 being moved from classes 2 and 4

Table 8.7 Means and sums of squares for cars classification

	Beauty	Comfort	Consumption	Power	Price	Reliability	Safety
Class 2							
Means	1	0.75	0.25	0.5	0.5	0.5	0.5
Sums of Squares	0	0.75	0.75	1	1	1	1
Class 3							
Means	0.67	1.00	1.00	0.33	1.00	0.67	0.33
Sums of Squares	0.67	0.00	0.00	0.67	0.00	0.67	0.67
Class 4							
Means	0.70	0.40	0.20	0.70	0.40	1.00	1.00
Sums of Squares	2.10	2.40	1.60	2.10	2.40	0.00	0.00
Class 5							
Means	0.33	1	1	0.33	0.33	1	1
Sums of Squares	0.67	0	0	0.67	0.67	0	0
Global means	0.7	0.65	0.45	0.55	0.5	0.85	0.8
Interclasses squares	0.77	1.40	2.60	0.52	0.93	0.88	1.53

Table 8.8 PseudoF and weighted PseudoF values for cars classification

Statistics	Triangular classification	Normal classification
PseudoF	2.22	1.84
WPseudoF	3.19	2.44

to Class 3. The weights for the computation of PseudoF are those employed to create the classifications.

Table 8.8 shows that, with a relatively larger between classes variance, the statistics present better values for the first classification.

References

Almeida-Dias, J., Figueira, J. R., & Roy, B. (2010). ELECTRE TRI-C: a multiple criteria sorting method based on characteristic reference actions. *European Journal of Operational Research, 204*(3), 565–580.

Almeida-Dias, J., Figueira, J. R., & Roy, B. (2012). A multiple criteria sorting method where each category is characterized by several reference actions: the ELECTRE TRI-nC method. *European Journal of Operational Research, 217*(3), 567–579.

Calinski, T., & Harabasz, J. (1974). A Dendrite method for cluster analysis. *Communications in Statistics, 3*, 1–27.

Roy, B. (1968) Classement et choix en presence de points de vue multiples (la method ELECTRE), pp. 57–75, La Revue d'Informatique et de Recherche Opérationelle (RIRO).

Sant'Anna, A. P., Costa, H. G., & Pereira, V. (2012). CPP-TRI: um método de classificação ordenada baseado em composição probabilística. *Relatórios de Pesquisa em Engenharia de Produção, 12*(8), 104–117.

Yu, W. (1992). *ELECTRE TRI—Aspects methodologiques et guide d'utilisation.* Document du LAMSADE: Université de Paris-Dauphine, Paris.

Zadeh, L. (1965). Fuzzy Set. *Information Science, 8*(3), 338–353.

Chapter 9
Capacities Determination

Abstract A capacity may be derived from the observed evaluations by following the principle of maximizing posterior probabilities. By this choice, of the posterior probability as paradigm, will be assigned highest capacity to those sets of criteria for which is highest the probability of some alternative maximizing the preference. Thus more importance is assured to those criteria with highest power of isolated discriminating a best option and if criteria repeat each other their importance is not magnified by such repetition.

Keywords Capacity · Choquet integral · Estimation · Preference by at least one criterion · IDH

9.1 The Maximization Capacity

A capacity for a set of criteria is here derived from the probabilities of maximizing preference according to them. The finality of adopting such form of derivation is to assign highest capacity to those sets of criteria for which is highest the probability of some alternative maximizing the preference.

The rule to derive the capacity from the probabilities of preference maximization will consist of first computing, for each subset of the set of criteria, the probabilities of maximization according to at least one of the criteria of the subset. The capacity will be proportional to the maximum, along the alternatives, of these probabilities of preference maximization. Its final value will be obtained by proportionally rescaling this vector of maxima in such way to give the value 1 to the set of all the criteria.

Formally, the capacity estimation algorithm to generate the capacity of a subset $\{C_1, \ldots, C_s\}$ of s criteria has the two following steps:

First compute

$$P(\{C_1, \ldots, C_s\}) = \max_a(1 - (1 - P_{a1}) \ldots (1 - P_{as})),$$

A. Parracho Sant'Anna, *Probabilistic Composition of Preferences, Theory and Applications*, Decision Engineering,
DOI 10.1007/978-3-319-11277-0_9

for a varying along the set of alternatives and P_{aj} denoting the probability of alternative a maximizing the preference according to criterion j,
and finally compute

$$C(\{C_1, \ldots, C_s\}) = P(\{C_1, \ldots, C_s\})/P(U),$$

for U denoting the set of all the criteria.

Thus, for subsets of more than one criterion, the capacity will be proportional to the maximum, along the set of observed alternatives, of the complement to 1 of the product of the probabilities of not maximizing the preference according to each criterion in the subset.

By giving more importance in the composition of the preferences to the criteria or sets of criteria with higher probabilities of choice of some alternative, this rule guarantees the highest final preferences for the alternatives eventually chosen. This agrees with the principle of Bayesian statistical inference (Box and Tiao 1973) of assigning to the parameters a probability distribution that maximizes the posterior probability of observing whatever has been observed.

This algorithm assumes independence between the disturbances affecting the evaluations according to each criterion. The choice of the independence assumption has the advantage of maximizing the attention given, in the estimation process, to the numerical evaluation according to each criterion. It does not mean, as already pointed out in preceding chapters, assuming unsubstitutability of criteria.

9.2 Use of the Capacity to Evaluate the Alternatives

With this construction of the capacity, to compute the aggregate probability of maximizing the preference for a given alternative with respect to a set of criteria by the Choquet integral (Choquet 1953), the following procedure may be followed.

First, to the lowest between the probabilities of maximizing the preference for the alternative according to a single criterion, add the product of the difference between the second lowest of such probabilities and the lowest by the capacity derived from the maximum along all the alternatives of the probabilities of being of the best according to at least one of those criteria different from that for which the alternative presents its lowest probability. Then, to this sum add the product of the difference between the third and the second probabilities by the capacity derived from the maximum along all the alternatives of the probabilities of being of the best according to at least one of the criteria in the set complementary to that formed by those with the two lowest probabilities. And so on.

To make clearer the procedure, let us consider, for instance, the case of four criteria and let us denote by P_j the probability of the alternative being that one maximizing the probability of preference by the j-th criterion and by τ the permutation of $\{1, 2, 3, 4\}$ such that

$$P_{\tau(1)} \leq P_{\tau(2)} \leq P_{\tau(3)} \leq P_{\tau(4)}.$$

The fist summand will be $P_{a\tau(1)}$.

The second, the product of

$$P_{\tau(2)} - P_{\tau(1)}$$

by

$$\max_a \left(1 - \left(1 - P_{a\,\tau(2)}\right)\left(1 - P_{a\,\tau(3)}\right)\left(1 - P_{a\,\tau(4)}\right)\right) \Big/ \max_a (1 - (1 - P_{a1})(1 - P_{a2})(1 - P_{a3})(1 - P_{a4})).$$

The third, the product of

$$P_{\tau(3)} - P_{\tau(2)}$$

by

$$\max_a \left(1 - \left(1 - P_{a\,\tau(3)}\right)\left(1 - P_{a\,\tau(4)}\right)\right) \Big/ \max_a (1 - (1 - P_{a1})(1 - P_{a2})(1 - P_{a3})(1 - P_{a4})).$$

Finally, the fourth will be the product of

$$P_{\tau(4)} - P_{\tau(3)}$$

by

$$\max_a \left(P_{a\tau(4)}\right) \Big/ \max_a (1 - (1 - P_{a1})(1 - P_{a2})(1 - P_{a3})(1 - P_{a4})).$$

9.3 Example of Capacity Estimation

Here will be studied again the problem of determining capacities for the construction of an aggregate index of sustainable management of resources or of sustainable development of regions or countries.

An index of sustainable development must compose evaluations of environmental, social and economic risks (Stiglitz et al. 2009). As seen in the preceding chapters, to cover all the necessary dimensions, the indicators composed may intercept and some of them may be substitutable by others in the set considered.

The criteria will be the same considered in the preceding chapters. The alternatives will be the 187 countries and the initial probabilities of maximizing preference for each alternative according to isolated criteria will be those there considered.

The capacity derived from the probabilities of presenting the best scores on the components of HDI assuming triangular distributions is given in Table 9.1, for L denoting longevity, MS denoting mean schooling, ES denoting expected schooling and I denoting income.

If the decision maker desires to pay more attention to the small values of the attributes, the probabilities of minimizing may be used instead of the probabilities of maximizing. The capacity derived from the probabilities of minimizing, shown in Table 9.2, gives a much lower importance to income.

Table 9.3 presents the scores for the same 30 countries analyzed in preceding chapters, obtained combining the probabilities of maximizing and minimizing preference by the Choquet integral with respect to the respective capacities.

As the clearest extremes for the probability of maximizing are given by the income component, the countries with best performance on this feature will reach the best global score by the first approach. Thus, employing to combine the probabilities of maximization the capacity in Table 9.1, the 17 first countries of the list of 30 are those occupying the 17 first positions among the 187. Specially noticeable are the positions of Qatar and Lichtenstein as the first and the third. The evaluation based on the probabilities of minimization is much closer to those obtained in the previous chapter and to the ranking by HDI.

Table 9.1 Capacity for IDH components derived from the probabilities of maximizing

Criterion	Importance	Criteria	Importance	Criteria	Importance
L	0.3084	L or MS	0.5141	L or MS or ES	0.8204
MS	0.3887	L or ES	0.6098	L or MS or I	0.9334
ES	0.4304	L or I	0.8727	L or ES or I	0.9395
I	0.7533	MS or ES	0.6555	MS or ES or I	0.8816
		MS or I	0.8145		
		ES or I	0.8206		

Table 9.2 Capacity for IDH components derived from the probabilities of minimizing

Criterion	Importance	Criteria	Importance	Criteria	Importance
L	0.4459	L or MS	0.674	L or MS or ES	0.8624
MS	0.4251	L or ES	0.6013	L or MS or I	0.8103
ES	0.4603	L or I	0.5857	L or ES or I	0.7393
I	0.1534	MS or ES	0.7205	MS or ES or I	0.8504
		MS or I	0.5636		
		ES or I	0.5860		

Table 9.3 Scores by new capacities

IDH	Country	Maximization		Minimization	
Rank		Score	Rank	Score	Rank
1	Norway	0.009504	4	0.002451	2
2	Australia	0.01159	2	0.002493	3
3	United States	0.008827	5	0.002509	4
4	Netherlands	0.007694	11	0.002638	7
5	Germany	0.007587	12	0.002631	6
6	New Zealand	0.008155	8	0.002558	5
7	Sweden	0.007713	10	0.002648	8
8	Switzerland	0.008219	6	0.002365	1
9	Japan	0.008156	7	0.002658	9
10	Hong Kong	0.008106	9	0.002675	11
11	Iceland	0.007401	14	0.002674	10
12	Israel	0.007065	16	0.002744	12
13	Singapore	0.007455	13	0.002799	14
14	France	0.007115	15	0.002746	13
15	Liechtenstein	0.01149	3	0.002955	15
16	Qatar	0.012439	1	0.003249	16
17	Kuwait	0.005433	17	0.003521	17
18	Equatorial Guinea	0.003134	18	0.007462	18
19	Papua New Guinea	0.002961	19	0.008181	19
20	Djibouti	0.002834	21	0.009142	20
21	Sudan	0.002797	22	0.011429	23
22	Sierra Leone	0.002702	27	0.0122	26
23	C. African Rep.	0.002702	26	0.011566	24
24	Eritrea	0.002857	20	0.011027	22
25	Mali	0.002702	28	0.011732	25
26	Burkina Faso	0.00273	25	0.013164	27
27	Chad	0.002668	29	0.013946	29
28	Mozambique	0.002773	23	0.013678	28
29	D. R. Congo	0.002763	24	0.010653	21
30	Niger	0.00259	30	0.014862	30

References

Box, G. E. P., & Tiao, G. C. (1973). *Bayesian inference in statistical analysis*. Reading-MA: Addison-Wesley.

Choquet, G. (1953). Theory of capacities. *Annales de l'Institut Fourier, 5*, 131–295.

Stiglitz, J. E., Sen, A., & Fitoussi, J. P. (2009). Report by the Commission on the Measurement of Economic Performance and Social Progress. Retrived from http://www.stiglitz-sen-fitoussi.fr.

Chapter 10
Rough Sets Modeling

Abstract The transformation in probabilities of being the best can reduce the number of possible values of the attributes. By this property, it may be used to amplify the roughness of decision attributes in Rough Sets Theory applications. This can be explored to increase the index of quality of approximation and simplify the classification rules.

Keywords Rough sets · Reducts · Index of quality of approximation · Dominance · Probabilistic transformation

10.1 Roughness Modeling

Rough Sets Theory—RST (Pawlak 1982) is based on identifying approximately classes determined by a set of attributes, named decision attributes, according to another set of attributes, named condition attributes.

An important stage in the characterization of rough sets is the identification of reducts, subsets of the set of condition attributes able to offer the same quality of approximation as the whole set. The possibility of approximation and, consequently, the number of reducts depend on the roughness of the sets, a concept whose characterization has been centered, in previous developments, on properties of the vector of values of the different attributes, but may consider also the precision in the measurement of each attribute.

The approach here employed to simplify the approximation applies to the situation in which the attributes are ordered variables. For this situation was developed the extension of RST called by Greco et al. (2001, 2002) Dominance-based Rough Sets Approach—DRSA.

A form of easing the approximation on DRSA employing relaxation of the rules for the entry of alternatives in the approximation by ignoring some contradictions is Dominance-based Rough Sets with Variable Consistency VC—DRSA, developed by Greco et al. (2005).

A. Parracho Sant'Anna, *Probabilistic Composition of Preferences, Theory and Applications*, Decision Engineering,
DOI 10.1007/978-3-319-11277-0_10

New proposals in the same direction were presented in Sant'Anna and Moreira Filho (2013) based on joining neighboring classes. While VC—DRSA and variants of it are based on the size of the classes involved, this last approach is based on the distance between the values of the decision attribute in the two classes. To evaluate such distance the probabilistic transformation is used.

The idea behind this probabilistic approach is to reduce the numerical precision in identifying according to the decision attributes by reducing the number of different possible values for them. This is made possible by the transformation of the vector of initial observations of the attribute into a vector of probabilities of presenting the highest (or lowest) value, because many alternatives have a very low probability of being the best (or the worst).

In this development is considered the most frequent case, of a unique decision attribute. The extension of the methods proposed for the case of more than one decision attribute is simple, once a dominance relationship is established in accordance with the set of all decision attributes together. To this end one can also use the composition of probabilistic preferences.

10.2 Rough Sets Theory

If two alternatives have the same values for all condition attributes, they are considered indiscernible. It counts as an inconsistency two alternatives classified as indiscernible having different classifications according to the decision attributes.

For each set of condition attributes P and each alternative x of the universe U of alternatives to be classified, denote by P(x) the set of alternatives indiscernible of x according to P:

$$P(x) = \{y \in U/x \textit{ and } y \textit{ are indiscernible by } P\}.$$

For every subset X of U and every set of condition attributes P are defined two approximations: the lower approximation of X by P and the upper approximation of X by P, defined by

$$\underline{P}(X) = \{x \in U \,| P(x) \subseteq X\}$$

and

$$\overline{P}(X) = \{x \in U \mid P(x) \cap X \neq 0\}.$$

It is easy to see that

$$\underline{P}(X) \subseteq \mathrm{X} \subseteq \overline{P}(X).$$

The boundary of X according to P is defined by

$$F_P(X) = \overline{P}(X) - \underline{P}(X),$$

Pawlak (1982) defines a set X as rough if its boundary is not empty and as crisp if it is empty.

Pawlak (1991) defines an index of quality of approximation γ by the set of condition attributes P for the set of decision attributes D as the proportion of alternatives whose classification according to D is not contradicted by indiscernibilities according to P. γ can be determined complementarily by the proportion of alternatives in boundaries of the sets of the partition $\{X_i\}_{i=1,\ldots,n}$ of U determined by D. Formally,

$$\gamma_P(D) = 1 - \sum_i \frac{|F_P(X_i)|}{|U|}.$$

The goal of RST is to establish decision rules in terms of values of the condition attributes to inform on the relevant classes determined by the decision attributes. By a principle of parsimony, rules employing less condition attributes are preferable.

If a set of condition attributes P has the same index of quality of approximation of a larger one, i.e., if $\gamma_P(D) = \gamma_Q(D)$ for P ⊆ Q, this means that there is no preference between two alternatives x and y according to attributes in Q\P that might erase any indiscernibility increasing a boundary according to P. If that happens, decision rules employing the attributes in Q can be replaced by rules employing only the attributes in P.

Pawlak (1991) defines as a reduct a set of condition attributes P such that any subset of P has a smaller quality of approximation than P. The smaller the number of attributes in the reducts, the simpler the decision rules.

If the attributes determine dominance relationships, i.e., if each of them determines an order relation in U, it is easier to measure the quality of approximation. The adaptation of RST for the case of dominance as set forth below was developed by Greco et al. (2001).

Suppose the classes of the partition $\{Cl_t\}_{t=1,\ldots,n}$ determined by the decision attributes ordered in such way that, if $r > s$, the alternatives in Cl_r are preferable to those in Cl_s. Then, it is enough to deal with the cumulative classes

$$Cl_t^{\geq} = \bigcup_{s \geq t} Cl_s \quad or \quad Cl_t^{\leq} = \bigcup_{s \leq t} Cl_s; \quad t = 1, \ldots, n.$$

For the condition attributes, x dominates y with respect to P, or “x P-dominates y”, or y is “P-dominated by x” or “xD_Py”, if the preference for x is higher or equal than the preference for y according to all the elements of P.

For dominance relations, it is interesting to consider the set of alternatives dominating a given alternative x:

$$D_P^+(x) = \{y \in U : yD_Px\},$$

as well as the set of alternatives dominated by x:

$$D_P^-(x) = \{y \in U : xD_Py\}.$$

Once defined the classes and the entire list of the P-dominating and P-dominated sets, it is possible to define the upper and lower approximations. Regarding the definitions above, the difference is that now, dominance relationships are used instead of indiscernibility.

The lower and upper approximations can be set respectively for the upper cumulative classes by:

$$\underline{P}(Cl_t^{\geq}) = \{x \in U : D_P^+(x) \subseteq Cl_t^{\geq}\}$$

and

$$\overline{P}(Cl_t^{\geq}) = \{x \in U : D_P^-(x) \cap Cl_t^{\geq} \neq \varnothing\}.$$

Analogously for the lower cumulative classes:

$$\underline{P}(Cl_t^{\leq}) = \{x \in U : D_P^-(x) \subseteq Cl_t^{\leq}\}$$

and

$$\overline{P}(Cl_t^{\leq}) = \{x \in U : D_P^+(x) \cap Cl_t^{\leq} \neq \varnothing\}.$$

Boundaries and the index of quality of approximation can be defined as before, being enough consider in the computation of the index of quality of approximation the boundaries of cumulative classes. Likewise, it is enough formulating the decision rules in terms of pertinence to cumulative classes.

10.3 VC-DRSA

VC-DRSA is based on a relaxation of the conditions for entrance in the approximations, controlled by a level of consistency l, which can vary from zero to one. In the case of a lower class $Cl_t^{\leq}$, for an alternative x be included in the lower approximation, the requirement of no alternative outside $Cl_t^{\leq}$ being dominated by x is relaxed and some alternatives outside $Cl_t^{\leq}$ may be dominated by x, as long as the proportion of those not belonging to $Cl_t^{\leq}$ does not exceed 1-l. Similarly, for an upper class, the lower approximation of consistency level l is:

$$\underline{P}^l(Cl_t^{\geq}) = \left\{ x \in Cl_t^{\geq} : \frac{|Dp^{+}(x) \cap Cl_t^{\geq}|}{|Dp^{+}(x)|} \geq l \right\}$$

So, if the consistency level is fixed at $l = 1$, the result of application of VC-DRSA will be the same as that of DRSA, but, for $l < 1$, more elements may belong to the approximations and the boundaries may become smaller.

A difficulty with this rule is that it has the effect of, for instance, leading some alternative x to be accepted in the lower approximation of a lower class while other alternative, dominated by x, is not accepted because it dominates less alternatives. This is not reasonable, as being dominated by x and not dominating other alternatives should be a stronger reason for this second alternative to belong to the lower approximation of the lower class.

Changing the condition of being dominated by that of dominating, identical inconsistencies may occur in the higher approximations of the upper classes. Several alternatives (Blaszczynski et al. 2006, 2009; Inuiguchi and Yoshioka 2006; Inuiguchi et al. 2009; Deng et al. 2011) have been proposed to avoid this contradiction, but all preserving the principle of establishing the level of consistency in terms of proportion of inconsistencies, disregarding the distance involved in the dominance.

10.4 Aggregation of Classes

In Sant'Anna and Moreira Filho (2013) is proposed the strategy of changing the original data by augmenting the roughness in the measurements to avoid contradictions that may increase the number of alternatives in the boundaries. The basic idea of this strategy is to move alternatives in small classes with extreme values to neighboring classes, by replacing such extreme values by approximations.

Increasing roughness in this way identifies alternatives with close values in the decision attribute. With the values of the condition attributes unchanged this will result necessarily in the possibility of reducing the number of contradictions, increasing the quality of approximation and reducing the size of the reducts.

In fact, the quality of the approximation is increased whenever the cardinality of the boundaries decreases and what reduces the boundaries is reducing, for each alternative x, the number of alternatives indiscernible from x and located outside the class in which the decision attribute locates x. If the values of the condition attributes are not changed, the indiscernible objects remain the same. Thus, the increase in the size of the classes determined by the decision attribute is forcefully associated to reduction of the number of indistinguishable alternatives outside.

The shrinking of the reducts can be proved in two steps. First, it is easy to see that no reduction is lost when joining the classes determined by the decision

Table 10.1 Data aggregation

Alternatives	D	P_1	P_2	D_0
A	2	1	1	3
B	3	1	2	3
C	5	2	3	5
D	6	3	3	6
E	7	4	4	7
F	8	5	5	8

attribute. Indeed, if a set P_1 of condition attributes contained in the set of attributes P is a reduct, then no indiscernibility according to attributes in P/P_1 is able to withdraw alternatives from any boundary. With the expansion of the classes determined by the decision attributes there is no way to raise the possibility of occurrence of any such withdrawal. Thus the size of the reducts cannot increase.

On the other hand, joining classes opens the possibility of appearing new reducts of smaller size. Indeed, if, before aggregation, P was a reduct and $P_1 \subseteq P$ was not, this is necessarily due to the existence of at least one indiscernible pair of alternatives (x, y) according to some element of P/P_1 that would place such alternatives in a boundary situation. With the aggregation of classes, such alternatives may become members of the same class.

The following example demonstrates concretely how the quality of the approximation may increase and the size of the reducts decrease with the union of classes.

In Table 10.1, D_0 results from joining two classes at the lower end of the range of values of D. With D as the decision attribute, P_1 and P_2 are needed to achieve the maximum quality of approximation of 1. But replacing D by D_0, it is easy to see that P_1 alone offers this quality of approximation of 1.

10.5 A Probabilistic Aggregation Algorithm

In this strategy, the values of the decision attribute are treated as the result of distortion of unknown values by random disturbances. This allows, as in preceding chapters, for replacing them by the probabilities of presenting an extreme value. The probability of presenting the highest value for two alternatives with the same evaluation will be the same and will be lower than that of an alternative belonging to a higher class. This argument is reversed when the transformation is based on the probability of presenting the lowest value.

What is explored to increase the index of quality of approximation is the fact that the probability of maximizing (or minimizing) becomes so close to zero for alternatives in extreme classes that, with a suitable decimal approximation, values in neighbor classes may be considered equal.

As in VC-DRSA (where the index l for input objects in the lower approximation is not fixed), an approximation threshold can be determined according to the circumstances. The limit to the approximation, denoted by l^* may be changed by the analyst.

As in VC-DRSA, a very low value of l^* may result in an excessive relaxation for entering values in the approximation classes. On the other hand, a high value for l^* allows for the union of classes that are not classes of rare alternatives in the tail of the classification.

As a rule, the approximation to the third decimal enables a satisfactory number of unions. This corresponds to consider equal those evaluations differing by less than 0.0005. With the probabilistic transformation, this means equaling to zero all evaluations of the probability of being the best below 0.0005.

If the important distinctions are between the alternatives presenting large values, the small probabilities will be assigned to the alternatives with low values for the decision attribute. Similarly, if the important distinctions are between the alternatives presenting large values, the small probabilities that will be considered equal will appear in the classes with high evaluations.

Merging classes by performing changes only in the decision attribute, no premise for application of DRSA is violated. Only classes determined by the decision attribute are pasted. Thus, the reduction of classes by the probabilistic transformation does not prevent the subsequent enforcement of any method of extraction of rules or imposition of consistency and the method for reduction of the number of classes can be applied in conjunction with any of the techniques developed to improve the quality of the approximation starting from VC-DRSA.

A combination of RST with the probabilistic transformation was also performed in the reverse direction in Sant'Anna (2004).

10.6 Example of Car Models

Consider the problem of choice among 20 car models. Assume that we intend to explain the decision attribute D in Table 10.2 by the condition attributes C_1 and C_2.

The quality of approximation is 0.4, as eight of the 20 alternatives are consistently classified: Car1, Car2, Car3, Car4, Car1, Car12, Car16 and Car17.

Applying the transformation of the decision variable into probabilities of preference (assuming a normal distribution with variance estimated by the observed variance) and rounding to three decimal places, the decision attribute receives the values in the last column. Thus, Car18, Car19 and Car20 are joined in the same decision class, with the value of 0 for the decision attribute. With this reduction in the number of classes, these three alternatives become consistently classified and the quality of approximation increases to 0.55.

Table 10.2 Cars evaluations

Cars	C1	C2	D	D_0
Car1	5	6	5	0.373
Car2	4	5	4	0.071
Car3	4	0	4	0.071
Car4	4	0	4	0.071
Car5	3	0	4	0.071
Car6	3	0	4	0.071
Car7	0	4	4	0.071
Car8	0	3	4	0.071
Car9	0	2	4	0.071
Car10	3	0	3	0.007
Car11	2	0	3	0.007
Car12	2	0	3	0.007
Car13	0	4	3	0.007
Car14	0	2	3	0.007
Car15	0	2	3	0.007
Car16	0	1	3	0.007
Car17	0	1	3	0.007
Car18	0	0	2	0
Car19	0	0	2	0
Car20	0	0	1	0

References

Blaszczynski, J., Greco, S., Slowinski, R., & Szelag, M. (2006). On variable consistency dominance-based rough set approaches. In *Lecture Notes in Artificial Intelligence* (vol. 4259, pp. 191–202). Berlin: Springer.

Blaszczynski, J., Greco, S., Slowinski, R., & Szelag, M. (2009). Monotonic variable consistency rough set approaches. *International Journal of Approximate Reasoning, 50*(7), 979–999.

Deng, W., Wang, G., Yang, S., & Hu, F. (2011). A new method for inconsistent multicriteria classification. In *Proceedings of the 6th International Conference on Rough sets and Knowledge Technology* (pp. 600–609).

Greco, S., Matarazzo, B., & Słowiński, R. (2001). Rough sets theory for multi-criteria decision analysis. *European Journal of Operational Research, 129*(1), 1–47.

Greco, S., Matarazzo, B., & Słowiński, R. (2002). Multicriteria classification by dominance-based rough set approach. In W. Kloesgen & J. Zytkow (Eds.), *Handbook of data mining and knowledge discovery*. New York: Oxford University Press.

Greco, S., Matarazzo, B., Słowiński, R., & Stefanowski, J. (2005). Variable consistency model of dominance-based rough set approach. In W. Ziarko & Y. Yao (Eds.), *Rough sets and current trends in computing, Lecture notes in artificial intelligence* (pp. 170–181). Berlin: Springer.

Inuiguchi, M., & Yoshioka, Y. (2006). *Variable-precision dominance-based rough set approach.* Annals of the 5th International Conference on Rough Sets, Kobe-Japan, pp. 203–213.

Inuiguchi, M., Yoshioka, Y., & Kusonoki, Y. (2009). Variable-precision dominance-based rough set approach and attribute reduction. *International Journal of Approximate Reasoning, 50*, 1199–1214.

Pawlak, Z. (1982). Rough sets. *International Journal of Computer and Information Science, 11*, 341–356.

Pawlak, Z. (1991). *Rough sets: Theoretical aspects of reasoning about data*. Berlin: Springer.

Sant'Anna, A. P. (2004). Rough sets in the probabilistic composition of preferences. In B. De Baets, R. D Caluwe, G. De Tré, J. Fodor, J. Kacprzyk & S. Zadrozny (Eds.), *Current issues in data and knowledge engineering, Proceedings of Eurofuse 2004*, pp. 407–414.

Sant'Anna, A. P., & Moreira Filho, R. M. (2013). Agregação automática de classes nos atributos de decisão em aplicações de Conjuntos aproximativos com dominância. *Sistemas & Gestão, 8*, 68–76.

Chapter 11
Application to FMEA Priority Assessments

Abstract A risk priority probability, obtained by multiplying the probabilities of being the mode of failure of higher risk simultaneously with respect to severity, occurrence and detectability, is here employed instead of the classical priority risk number of FMEA. A probabilistic classification of risks with respect to classes previously determined is also discussed.

Keywords FMEA · Risk priority number · Risk priority probability · Severity · Occurrence · Undetectability · Ordered classes of risk

11.1 Risk Priority Probabilities

A natural example of the composition of preferences by joint probabilities is the calculation of priorities of the failure modes in Failure Modes and Effects Analysis—FMEA (U.S. Defence Department 1949). An important feature of FMEA is the assignment of a risk priority number (RPN) to each failure mode. To determine that number, is assigned to each failure mode a score from 1 to 10, or from 1 to 5, with respect to three features: severity, frequency and detectability. The RPN is simply obtained by multiplying the values of the scores on these three features.

In Sant'Anna (2012) is proposed the replacement of the RPN by a risk priority probability (RPP), obtained by multiplying the probabilities of being the mode of failure of highest risk according to each of these criteria.

This multiplication may be justified in terms of probability theory by the combination of conditional probabilities. To evaluate a failure mode, we must first consider the likelihood of its cause. Conditionally on the occurrence of such cause, the evaluation of the risk in the mode of failure must then consider the probability of such occurrence not being detected in time to avoid its unwanted effects. Finally, conditionally on occurrence and undetectability is that the probability of an undesirable effect should enter the computation.

A. Parracho Sant'Anna, *Probabilistic Composition of Preferences, Theory and Applications*, Decision Engineering,
DOI 10.1007/978-3-319-11277-0_11

In each of these steps, the comparative assessment in terms of probability of being the most severe, the least detectable and the most frequent can be applied under the conditional framework. Thus, a natural measure of risk is the product of the probabilities of maximizing frequency by the conditional probability of maximizing undetectability given occurrence and the conditional probability of maximizing damage given undetectability and occurrence. Such multiplication corresponds to the pessimist and progressive probabilistic composition of the three criteria of FMEA.

Besides this theoretical reason, there is a practical reason to replace RPN by RPP. In the classical calculation of RPN, a change from 1 to 2 in the evaluation of frequency, for instance, corresponding to a change in the evaluation of the cause from extremely rare to a little less frequent, doubles the product of the three factors, so increasing the RPN much more than it would be reduced by a change from 10 to 9, from extremely common to a little less frequent. If the priority is determined by the probability of being at the same time the least detectable, most frequent and most serious, this distortion is limited, as the probability according to any criterion that the risk is the highest varies among alternatives of high risks much more than among alternatives of low risks.

To be able to perform the probabilistic transformation, one must model the distributions of the random disturbances. In Sant'Anna (2012) is set for FMEA with scores ranging from 1 to 10 a triangular distribution with bounds at 0 and 11. With the limits on the exact possible ends of 1 and 10, as well as with limits more distant, spaced by a small fraction of these lower and upper ends, for instance to 0.9 and 11, very close ranks have been obtained in practice.

On the other hand, by setting the ends of the distributions near the largest and the smallest observed values, instead of the maximum and minimum possible values of 1 and 10, different results are obtained. Approaching the bounds to the observed values corresponds to discarding the equal weighting of criteria implicit in the multiplication formula and adopting the weighting implicit in the variability of the evaluations given by experts.

A basic feature of FMEA is to let the experts the possibility of giving more weight to one or another criterion by placing a greater distance between the scores for the modes of failure according to the criterion that they believe to be more important than the others. It may happen, on the other hand, that it is the greater difficulty in assessing the importance of the modes of failure according to a criterion that leads the evaluators to shorten distances in the evaluations according to such criterion. If this is the case, setting the ends of the distribution next to the observed values, by increasing the importance of the factors where there is less variability, will unduly distort the composition.

Finally, instead of just ranking the modes of failure, experts on FMEA sometimes present the results of their evaluation in terms of classes of risk. The probabilistic approach allows for directly classifying the modes of failure by the trichotomic probabilistic measurements described in Chap. 8. For that, absolute profiles for the classes may be derived. Basic profiles for each class may be established with equal values for all the criteria.

11.2 FMEA

FMEA is a tool to improve reliability. The basic objective of FMEA is to eliminate potential causes of failures before they take place.

FMEA may be applied in the development of a product as well as in the modeling of a manufacturing process. It is preferably applied before beginning manufacture or assembly, but can be applied to processes in progress, on a cyclic movement to raise quality. Once developed a product or process whatsoever, it must be regularly reviewed, its output compared to that imagined, to the attention to the expected faults added the focus on those really happening in the day-to-day process management and product use.

Application of FMEA involves recording the possible failures and evaluating its severity, frequency and detectability objectively in a document, the form of FMEA. In the form of FMEA are initially recorded functions and main features of the product or process, modes of potential failures for each function; causes and effects of each mode of failure and current controls.

FMEA starts basically with the formation of a group of people to identify the product or process in question, its functions, the types of failures that can occur, the effects and the possible causes of each failure, to develop the form of FMEA. Then the risks for each cause of failure are evaluated by means of the three indices. Based on this assessment, are proposed actions able to reduce these risks by eliminating the most important causes of failure, those whose combined priority index exceeds a predetermined threshold, increasing in this way the reliability of the product or process.

At the stage of risk assessment, evaluations are recorded in the form of indices of severity (S), occurrence (O) and detectability (D) for each mode of failure, following previously defined patterns. The first of these indices measures the severity of the consequences attributed to an undetected occurrence of the failure. The second measures the probability of occurrence, i.e., the frequency of observation expected for the failure. The third measures the difficulty of detecting the failure in a timely manner. These indices are used in the traditional approach to calculate the RPN by multiplying their three values.

The identification of the values for these indices is in principle performed in meetings of the group of analysts. The development of communication tools in recent years favors, however, the use of isolated assessments that can be revised iteratively.

When the group is deciding on values according to one factor, the other indices cannot be taken into consideration, i.e., the evaluation of each index is independent. For example, when evaluating the level of severity of a particular effect, a low value should not be assigned to this index only because the probability of detection is high.

Further details and reviews of FMEA can be found in AIAG (2008) or McDermott et al. (1996), for instance.

11.3 Risk Indices

According to each of the criteria of severity, occurrence and detectability, the modes of failure are classified into a certain number of levels, properly determined previously. The previous description of levels based on the reality of the production process is one of the most important recommendations for FMEA (Stamatis 1995; Pfeifer 2002). Tables 11.1, 11.2 and 11.3 below provide an example of generic descriptions for five levels indices.

The construction of the tables of identification of the levels and the association to each mode of failure of a description in each table is an approximate process. Bad results in the application of the technique are frequently associated to errors in the construction and use of the scales. For this reason is important to take into account the uncertainty inherent to the evaluations.

Uncertainty is common in risk evaluations. It is expected that the available data are subject to fluctuations in their values if measurements are taken in different instances. When the preferences, as in the case of assigning priorities to possible

Table 11.1 Levels of severity

Level	Importance	Characterization
1	Very low	The use will not care
2	Low	Light loss of performance
3	Moderate	Real source of dissatisfaction
4	High	Inoperative system
5	Very high	Inoperative system menacing security

Table 11.2 Levels of undetectability

Level	Importance	Characterization
1	Very low	Certainly will be detected
2	Low	Most possibly detected
3	Moderate	Median chance of passing undetected
4	High	Most possibly undetected
5	Very high	Certainly will not be detected

Table 11.3 Levels of occurrence

Level	Importance	Characterization
1	Very low	1 in 1.000.000
2	Low	1 in 10.000
3	Moderate	1 in 1.000
4	High	1 in 1.00
5	Very high	1 in 10

failures, do not come from unified processes, but from application of personal value systems, the treatment of uncertainty becomes even more difficult.

Modifications to FMEA practice have been proposed to account for uncertainty. Unsatisfactory results of the application of these proposals can be attributed to the direct form of composition of the specific indices rather than to the account of the uncertainty taken in the calculation of these specific indices (Jin 2000; Lee 2001).

An often explored option is the use of fuzzy logic (Zadeh 1978) in the combination of the indices, whether or not these criteria are measured by fuzzy numbers (Bowles and Pelaez 1995; Pillay and Wang 2003). This form of composition results in prioritizing maximum or minimum values, disregarding the intermediate ones. The use of the product, on the contrary, by considering all the values, provides greater robustness to the final outcome of FMEA.

Other proposed changes are based on standardizing the values by dividing by the amplitude observed in each index, to correct for the different variability in the three components (Tay and Lim 2006; Keskin and Ozkan 2009). This may result in neglecting the information provided by this variability, allowing for distortions even more difficult to identify in the final outcome than those resulting from the direct multiplication of the original values.

The transformation into probabilities of preference and the probabilistic composition by the product of the probabilities provide greater security to handle the uncertainty in each factor and the differences in variability in the evaluations of the different risk factors.

11.4 Practical Considerations

An example of modeling uncertainty is given by the analysis in Sant'Anna et al. (2014) of the data of Chuang (2010) about 23 modes of failures in the services of a hypermarket. That is a typical case of the service sector, where scales of equally spaced values are employed with the identification of the importance levels in the scale left open to the evaluators.

The importance of the modes of failure according to the three FMEA criteria was determined in the study of Chuang (2010) by averages of evaluations by a team of 100 specialists. This approach allows for statistically estimating the dispersion in the evaluations according to each criterion. For instance, an estimator for the range of the distributions related to severity may be derived from the vector of 23 sample ranges observed in the 23 sets of 100 evaluations of each mode of failure according to severity.

Alternatively, the information on the dispersion within the samples of 100 evaluations may be left out of consideration to employ, instead, the variation of the means of such samples along the 23 modes of failure. It can be argued that the variation along the set of specialists does not correctly inform on the uncertainty in the application of the criterion, as all of them may be not equally affected by this uncertainty. The variability observed along the evaluations of the different modes of failure—i.e., the variability inter failures—may inform more on the variation on the evaluation of each failure—the variability intra failure—than the outer information brought by the variation between specialists.

11.5 Classification Example

A comparison of results of probabilistic composition with other approaches, based on data of Tay and Lim (2006), is performed in Sant'Anna (2012). The modes of failure are classified according to RPP into the five classes of Table 11.4 with extremes derived from the number of modes of failure examined.

In Sant'Anna (2012), the combination of this automatic classification by global probabilistic evaluations with another directly provided by experts is studied.

The data collected by Tay and Lim (2006) refer to three processes in semiconductors production. The first is the test handler process. It admits failures in handling integrated circuits for testing through an interface unit. Table 11.5 presents for this process the initial evaluations according to each factor of risk and the results of the probabilistic composition. The fourth column presents the probabilities of being the highest priority according to the three criteria together multiplied by 10^6. The fifth column presents the result of application of the thresholds in Table 11.5 to discretize the values in the fourth column. The last column presents the minimum between the values of the preceding column and the values assigned by experts.

For the three processes, the probabilistic classification is, in general, in lower levels than the experts' final classification, but the differences are small. Only once the experts reduce the probabilistic level, even so only from "very high" to "high", what would keep the corresponding failure still indicated to provoke corrective action in the process.

The failure object of reduction by the experts has values 9, 3 and 1 for severity, frequency and undetectability, respectively. The rank inversion in the experts' global evaluation was found to reflect a predilection for the application of the occurrence criterion.

The discretization preserves the variability of the global probabilities, with the classification of the modes of failure varying from 1 to 5 in all the three cases. The discrete probabilistic classification is much more spread than that of the global scores provided by the experts when asked to evaluate the whole set of potential failures.

Table 11.4 Probabilistic classes

$\leq n^{-3}/3$	$>n^{-3}/3$ and $\leq 2n^{-3}/3$	$>2n^{-3}/3$ and $\leq n^{-3}$	$>n^{-3}$ and $\leq 3n^{-3}/2$	$>3n^{-3}/2$
Very low risk	Low risk	Moderate risk	High risk	Very High risk

Table 11.5 Process failures probabilistic evaluations

Severity	Occurrence	Detectability	Risk	Discretized risk	Combined risk
1	1	1	10	1	1
1	1	2	12	1	1
1	1	3	13	1	1
1	2	3	15	1	1
1	3	1	13	1	1
2	1	1	12	1	1
3	1	1	13	1	1
4	1	1	15	2	2
2	1	2	13	1	1
1	5	1	18	2	2
5	1	1	18	2	2
3	1	2	15	1	1
2	1	3	15	1	1
2	2	1	13	1	1
3	1	3	17	2	2
1	7	1	27	2	2
7	1	1	27	2	2
3	2	1	15	1	1
5	2	1	20	2	2
3	3	1	17	2	2
7	3	1	35	3	3
9	1	1	56	4	4
9	2	1	63	4	4
9	3	1	71	5	4
2	9	1	62	4	4
7	8	1	97	5	5
3	9	1	71	5	5
7	10	10	1817	5	5

It can be noticed in Table 11.5 that the probabilistic priority of the failure with score 4 for severity and 1 for the other two criteria is higher than that of the failures with scores 1, 2 and 3 on different criteria. This is a consequence of the probabilistic composition taking into account the smaller variation of the vector of evaluations according to severity. This smaller variation makes the highest values with respect to severity more important.

In general, in this study, probabilistic composition led to results close to those of classic FMEA, but with important divergences. This may be seen as an assurance that the improvement resulting from taking into account random variability when combining the evaluations can be applied without risk of large deviations from the traditional practice.

In addition, the proposal to combine the probabilistic composition with the evaluation by experts and place the failure mode in the lower of the two classes appointed encourages the experts to center attention on the potential failures classified in high priority levels by the probabilistic composition. In those grounds, the experts can decide more attentively on the need of effectively changing the process under evaluation.

References

AIAG (2008). *Potential failure mode and effects analysis, Automotive industry action group*, 4th. Ed Southfield, MI.

Bowles, J. B., & Pelaez, C. E. (1995). Fuzzy logic prioritization of failures in a system failure mode, effects and criticality analysis. *Reliability Engineering & System Safety, 2*, 203–213.

Chuang, P. T. (2010). Incorporating disservice analysis to enhance perceived service quality. *Industrial Management & Data Systems, 110*, 368–391.

Jin, Y. C. (2000). Fuzzy modeling of high-dimensional systems: complexity reduction and interpretability improvement. IEEE Transactions on Fuzzy Systems, *2* (8), 212–221.

Keskin, G. A., & Ozkan, C. (2009). An alternative evaluation of FMEA: fuzzy art algorithm. *Quality & Reliability Engineering International, 25*, 647–666.

Lee, B. H. (2001). Using Bayes belief networks in industrial FMEA modeling and analysis, In *Proceedings of the reliability and maintainability symposium* (pp. 7–15).

McDermott, R. E., Mikulak, R. J., & Beauregard, M. R. (1996). *The basics of FMEA*. Portland, OR: Resources Engineering.

Pillay, A., & Wang, J. (2003). Modified failure mode and effects analysis using approximate reasoning. *Reliability Engineering & System Safety, 79*, 69–85.

Pfeiffer, T. (2002). *Quality management: strategies, methods, techniques* (3rd ed.), München: Hanser Fachbuchverlag.

Sant'Anna, A. P. (2012). Probabilistic priority numbers for failure modes and effects analysis. *International Journal of Quality & Reliability Management, 29*(3), 349–362.

Sant'Anna, A. P., Ribeiro, R. O. A., & Angulo Meza, L. (2014). Probabilistic composition in quality management in the retail trade sector. *International Journal of Quality & Reliability Management, 31*(6), 718–736.

Stamatis, D. H. (1995). *Failure modes and effects analysis*. Milwaukee: ASQC Quality Press.

Tay, K. M., & Lim, C. P. (2006). Fuzzy FMEA with a guided rules reduction system for prioritization of failures. *International Journal of Quality & Reliability Management, 23*(8), 1047–1066.

US Defence Department (1949). *Procedures for performing a failure mode, effects and criticality analysis*. Washington, D.C: US Government Printing Office, US Military Procedure MIL STD-1629.

Zadeh, L. A. (1978). Fuzzy sets as the basis for a Theory of Possibility. *Fuzzy Sets and Systems, 1*, 3–28.

Appendix
Basic Notions of Probability Theory

Abstract In this Appendix, the basic concepts of Probability Theory employed in the book are explained and their main properties are discussed.

Keywords Probability, Expected value, Additivity, Variance, Covariance, Independence, Conditional distribution

A.1 The Concept of Probability

A probability distribution, or, succinctly, a probability, on a finite set S, is an additive function p, from the set of parts of S, **P**(S), to the interval [0, 1], satisfying p(S) = 1. Additivity in a function whose arguments are sets means that, for any pair of disjoint sets A_1 and A_2,

$$p(A_1 \cup A_2) = p(A_1) + p(A_2).$$

In a reference to statistics practice, the sets that enter as arguments of a probability function are called events, the elements of S are called outcomes and S is called a sample space.

This concept may be extended from finite sets to any set S, but the set of events must present certain properties that sometimes may not be presented by the set **P**(S) of all parts of S. In the general definition, the set of events may be any nonempty subset **Λ(S)** of **P**(S) closed for complements and countable unions (a set of events satisfying these probabilities is called a sigma algebra). More precisely, a sigma algebra is any subset **Λ(S)** of **P**(S) satisfying: (1) if A ∈ **Λ(S)** then S\A ∈ **Λ(S)** and (2) if A_i ∈ **Λ(S)** for every set A_i of the sequence of sets $\{A_i\}_{i \in N}$, then $\cup_{i \in N} A_i$ ∈ **Λ(S)**.

A function p with domain **Λ(S)** and satisfying p(S) = 1 is then a probability in (S, **Λ(S)**) if and only if, for every sequence $\{A_i\}_{i \in N}$ of elements of **Λ(S)** satisfying $A_i \cap A_j = \Phi$ if i ≠ j,

$$p(\cup_{i \in N} A_i) = \sum_{i \in N} p(A_i).$$

A. Parracho Sant'Anna, *Probabilistic Composition of Preferences, Theory and Applications*, Decision Engineering,
DOI 10.1007/978-3-319-11277-0

So, the concept of probability involves three elements: the sample space S, the set of events **Λ(S)** and the probability function p. A triple (S, Λ, p), where Λ is a sigma algebra of parts of S and p is a probability with domain Λ is called a probability space.

Some properties follow immediately from the definition of probability. For instance,

$$p(\emptyset) = 0, \text{for any event A;}$$

denoting by A^C the complement of A with respect to S,

$$p(A) = 1 - p(A^c);$$

for any two events A and B,

$$p\ (A \cap B) = p\ (A) - p\ (A \backslash B),$$
$$A \subseteq B \rightarrow p\ (A) \leq p\ (B)$$

and

$$p(AUB) = p(A \backslash B) + p(A \cap B) + p\ (B \backslash A).$$

For every probability p on (S, Λ(S)) and any event B of Λ(S) with $p(B) > 0$, it is possible to derive another probability on S that will coincide with p if and only if $p(B) = 1$. This new probability is called the probability p conditional on B. This probability is denoted p(|B) and the probability of any event A of Λ(S) by p(|B) is given by

$$p(A|B) = p(A \cap B)/p(B).$$

An important use of conditional probabilities involves conditioning separately on the elements of a partition. It employs the following property.

Total Probability Theorem

For any countable partition $\{B_i\}_{i \in N}$ of S with $B_i \in \Lambda(S)$ and $p(Bi) > 0$ for all i, and for any $A \in \Lambda(S)$,

$$p(A) = \sum_{i \in N} p(A|B_i)\ p(B_i).$$

Here the reader must remember that a partition of a set S is any collection of mutually excludent subsets of S whose union is S; sets are mutually excludent if and only if they are pairwise disjoint, what means that the intersection of any pair of them is empty.

Independence

With respect to a probability p for which $p(A) \neq 0 \neq p(B)$,

$$\text{two events A and B are independent} \leftrightarrow p(A|B) = P(A).$$

Intuitively, independence means that the occurrence of B does not change the probability of A occurring or not. An equivalent definition of independence of events, simpler but less intuitive is:

$$\text{A and B are independent} \leftrightarrow P(A \cap B) = P(A)P(B).$$

This definition has the advantage of being applicable also to pairs of events of null probability. By this extension, events of null probability are independent of any other.

It is useful to consider the question about the independence of the members of a partition. They are not independent of each other (except those of null probability, which are unimportant for the applications of the Total Probability Theorem). In fact two disjoint events of nonnull probability are never independent of each other, as their conditional probabilities are null.

Another basic theorem, the Bayes Theorem, is used to discover the probability of each element of a partition conditionally on an event A that may be found to have occurred when is known the probability of A conditional on the events of the partition.

Bayes Theorem

$$p(B|A) = p(A|B)\ p(B) / p(A),$$

for any pair of events A and B.

In this context, $p(B|A)$ is called the posterior probability of B and $p(B)$ is called its prior probability.

This result has important practical consequences for statistical inference. One of the most important is to call attention to the importance of correctly evaluating prior probabilities.

Consider, for instance, the case of an event B of low prior probability such as an individual carrying the virus of a rare disease. That means, B is the set of members of a population S with the rare disease. Suppose a very accurate blood test is designed to detect the presence of such virus. Let A be the event that an individual in the population is pointed by this blood test as infected by the virus, that means, A is the set of the elements of S for which an application of the blood test gives a positive result.

To make things more concrete, let us suppose $p(A|B) = 0.99$, $p(A|B^c) = 0.001$ and $p(B) = 0.0001$. In that case, this prior being correct, you should not be very worried if the test points you as having the disease. Even though the probability of the test offering a wrong result is of only

$$\begin{aligned} p(A^c|B) * p(B) + p(A|B^C) * p(B^C) &= 0.01 * 0.0001 + 0.001 * 0.9999 \\ &= 0.0010009 \end{aligned}$$

and the probability of it presenting the positive result that you received is of only

$$\begin{aligned} p(A) = {} & p(A|B) * p(B) + p(A|B^C) * p(B^C) = 0.99 * 0.0001 + 0.001 * 0.9999 \\ = {} & 0.0010989. \end{aligned}$$

In fact, the posterior probability, although much higher than the prior, is still small:

$$p(B|A) = p(A|B) * p(B)/p(A) = 0.99 * 0.0001/0.0010989 = 0.09009.$$

Facing these results you may decide to change your prior to p(B) = 0.09 and take the blood test a second time. If the result is the same, then you should become worried. The posterior probability p(B|A) would now increase to around 0.99.

This underlines the importance of repetition to contest the prior, what was what Rev. Thomas Bayes emphasized when using his theorem, in the first half of the 18th century.

By combining Total Probability Theorem with Bayes Theorem posterior probabilities of all the elements of a partition can be obtained:

$$p(A_1|B) = [p(B|A_1)\ p(A_1)]/\sum_{i \in N} [p(B|A_j)\ p(A_j)].$$

Cumulative Distribution Functions

For the evaluation of preferences the most important attributes are evaluated by real numbers, and the most important events are numerical intervals. In that case the probabilities of preference will be defined in the sigma algebra of Borel, which is the set **B** of all countable unions of intervals and complements of intervals in the real line (that is, the sigma generated by the intervals—a sigma algebra generated by a given basis is the smallest sigma algebra containing it). To identify a probability p in this sigma algebra, it is enough to inform the probability of the intervals of the form $(-\alpha, x]$ for every $x \in \mathbf{R}$.

Conversely, by $p(-\alpha, x) = F(x)$, any rightcontinuous nondecreasing function F from R into [0, 1] satisfying $\lim_{x \to -\alpha} F(x) = 0$ and $\lim_{x \to +\alpha} F(x) = 1$ determines a unique probability p on (**R**, **B**). Functions of this kind are called cumulative probability functions or cumulative distribution functions (cdf).

Specially easy to use are the cdf which are not only rightcontinuous but absolutely continuous. Absolutely continuous are those functions F in the real domain for which there is another function in the real domain f such that, for all x, x, $F(x) = \int_{-\alpha}^{x} f(u)du$. f. f is called the density of F.

If F, absolutely continuous, is the cdf of p, the density f o F is determined by

$$f(x) = \lim_{\varepsilon \to 0}(p(x - \varepsilon,\ x + \varepsilon)/(2\varepsilon).$$

A.2 Random Variables

Functions employed to translate probabilities p in (S, Λ) to (R, B) are called random variables. A random variable in (S, Λ) is a function X with domain S such that the inverse image by X of any interval is an event in Λ.

For any random variable X, the probability p_X in (R, B) determined by

$$p_X((a, b]) = p(X^{-1}((a, b]))$$

is called the probability distribution of X and the cumulative probability function, F_X defined by

$$F_X(x) = p(X^{-1}((-\infty, x]))$$

is called the cumulative distribution of X. If F_X has a density, this is called also the density of X.

Two main types of random variables are relevant: those with probabilities p_X for which there is a countable set $S_p = \{x_i\}_{i \in N}$ of values for which $p_X(S_p) = 1$ and those with an absolutely continuous cdf. In the first type, named discrete, the probability distribution of X is determined by $\{p_X(x_i)\}_{i \in N}$. In the second, it is determined by its density.

To denote events in the Borel sigma algebra determined by values of a random variable X defined on a probability space (S, Λ, p) is employed a proper notation: for a phrase r(X) involving X, [r(X)] denotes the set {s∈S| r(X(s))} and p[f(X)] denotes the probability of the set of elements s of the sample space S for which the phrase f(X) is true when X is replaced by the real number X(s).

A.3 Expected Values

The expected value of a random variable X of the discrete type with $\sum_{i \in N} p_X(\{x_i\}) = 1$ is

$$E(X) = \sum_{i \in N} x_i p_X(x_i).$$

For a random variable X with density f_X, the expected value of X is

$$E(X) = \int x \, f_X(x)dx.$$

This sum and this integral may naturally not converge. So, for some variables, the expected value is not defined.

As a weighted average of the possible numerical values assumed by the random variable, with weights given by the probability, the expected value, like a centre of mass in a bar, is a number around which the probability distribution is spread. For this reason it is called a location parameter.

Other location parameters are, for continuous distributions, the median, that value m for which $F_X(m) = ½$, and the quantiles, those values x_q for which $F_X(x_q) = q$, and, for discrete distributions, the modes, those values M for which $p_X(M) \geq p_X(x)$ for any real x.

Similar to the location parameters, another kind of useful information about the distribution is given by the dispersion parameters, from which the most frequently used is the variance.

Variance of the random variable X is the nonnegative number

$$V(X) = E\left((X - EX)^2\right).$$

So the variance of X is the expected value of a measure of deviation of X from its location parameter EX.

From this definition follow that

$$V(X) = E(X^2) - (EX)^2.$$

The symmetric measure of deviation from EX employed in this definition is the square function. To bring the measurement to the same scale of X, instead of the variance, is used to measure dispersion the standard deviation, a parameter σ(X) defined as the square root of the variance. This means

$$\sigma(X) = E\left((X - EX)^2\right)^{1/2}.$$

A.4 Properties of the Expected Value and the Variance

The main property of the concept of expected value is linearity:

$$E(X + Y) = E(X) + E(Y)$$

and

$$E(cX) = cE(X) \text{ for any real c.}$$

For any event A of S and 1_A the random variable defined by

$$1A(s) = 1 \text{ if } s \in A \text{ and } 1_A(s) = 0 \text{ if } s \in S\backslash A,$$
$$E(1_A) = p(A).$$

By this correspondence, the concept of expected value may replace the concept of probability of an event.

Effects on the variance of translation and change of scale: for any real c,

$$V(X + c) = V(X)$$

and

$$V(cX) = c^2 VX.$$

A.5 Joint Distributions

The joint distribution of $X = (X_1, \ldots, X_n)$, a vector of random variables on (S, Λ, p), is the probability distribution p_X on $(\mathbf{R}^n, B^n)$ (where B^n denotes the sigma algebra generated by the products of intervals) determined by

$$p_X(I_1 x \ldots x I_n) = p(X_1 \in I_1, \ldots, X_n \in I_n),$$

for any set of intervals $(I_1, \ldots, I_n)$.

In the same way, the concepts of joint cumulative distribution function and of joint density extend the one-dimensional case. In the context of joint distributions of vectors X, the distribution of each one-dimensional random variable X_i, is called a marginal distribution. In the same way, its cdf is called a marginal cdf and its density is called a marginal density.

A vector of random variables $X = (X_1, \ldots, X_n)$ has a continuous distribution if and only if there is a positive function f_X such that the, for F_X the joint cumulative distribution function of X,

$$F_X(x_1, \ldots, x_n) = \int_{-\infty}^{x_1} \ldots \int_{-\infty}^{x_n} f_X(x_1 \ldots x_n) dx_1 \ldots dx_n$$

The following concepts, of correlation and independence between random variables, help to understand joint probability distributions.

Covariance of the pair of random variables X and Y is the expected value of the product of their deviations to the respective means:

$$\mathrm{Cov}(X, Y) = E[(X - EX)(Y - EY)]$$

So, if values above (or below) their expected values tend to occur together, then X and Y have a positive covariance. If values above the expected value for one of them tend to be accompanied by values below the expected value for the other then they have a negative covariance.

$$\mathrm{Cov}(X, Y) = E(XY) - E(X)E(Y)$$

and

$$V(X + Y) = V(X) + V(Y) + 2\mathrm{Cov}(X, Y)$$

To make possible comparison by controlling for the effect of scales, instead of covariance is employed the correlation coefficient.

The correlation coefficient between the random variables X and Y is

$$(X, Y) = \mathrm{Cov}(X, Y)/(\sigma(X)\sigma(Y))$$

The following are the properties that make the coefficient of correlation advantageous relatively to the covariance:

$$0 \leq |\rho(X, Y)| \leq 1$$
$$Y = cX \rightarrow |\rho(X, Y)| = 1 \text{ for } c{=}0,$$
$$\rho(X, Y) = +1 \text{ if } c > 0$$

and

$$\rho(X, Y) = -1 \text{ if } c < 0.$$

The random variables X and Y are independent $\leftrightarrow$

$$p[a \leq X \leq b, c \leq Y \leq d] = p[a \leq X \leq b]p[c \leq Y \leq d]$$

for every real numbers a, b, c and d.

For X and Y discrete random variables, this definition may be put in simpler terms:

$$X \text{ e } Y \text{ are independent} \leftrightarrow p[X = a, Y = b] = p[X = a]p[Y = b] \text{ for every real a and b.}$$

So the random variables 1_A e 1_B are independent $\leftrightarrow$ the events A and B are independent.

For continuous variables, X and Y are independent if and only if

$$f_{XY} = f_X f_Y.$$

X and Y are independent if and only if

$$E[g(X)h(Y)] = E[g(X)]E[h(Y)] \text{ for all real functions g and h,}$$

i.e., the expectation of the product of a function of X by a function of Y with respect to the joint distribution of X and Y is equal to the product of the expectations of the two random variables computed separately.

For continuous random variables X and Y with joint density f_{XY} and marginal densities f_X and f_y, this means:

$$\iint g(x)h(y)f_{XY}(x, y)dxdy = \int g(x)f_X(x)dx \int h(x)f_Y(y)dy.$$

So, X e Y independent $\rightarrow$

$$\begin{aligned} E(XY) &= E(X)E(Y), \\ \mathrm{Cov}(X, Y) &= 0, \\ \rho_{XY} &= 0 \end{aligned}$$

and

$$V(X + Y) = V(X) + V(Y).$$

A.6 Conditional Distributions

A more complete information on the dependence between random variables is given by the conditional expectations.

For any p and any random variables X and Y defined on (S, Λ) and any x ∈ **R** for which p[X=x] > 0, the expected value of Y with respect to p conditional on the event {s ∈ S | X(s) = x}, that means with respect to the probability p(|[X = x]) can be computed. It is denoted E(Y|X = x).

If X is a discrete random variable, a function $Y_{|X}$ with domain S and codomain **R** that for each s with p[X = X(s)] > 0 assigns $Y_{|X}(s)$ = E(Y|X = X(s)) is a random variable and its expected value $E(EY_{|X})$ equals the expected value EY of Y.

More generally, for any A ∈ Λ, the expected value of the restriction of $Y_{|X}$ to A coincides with the expected value of the restriction of Y to A, that means,

$$E\left(Y_{|X}1_A\right) = E(Y1_A).$$

Or extending yet a little more, for any random variable g(X, Y)

$$E(g(X, Y)) = E\left(g\left(X, EY_{|X}\right)\right).$$

This property is employed to extend the definition of conditional expectation. For any X and Y and any Z that is constant in any set of **Λ** where X is constant and satisfies E(g(X, Z)) = E(g(X, Y)), Z is a conditional expectation of Y given X.

This can be extended to conditioning on a vector of random variables. It is enough to replace in the above formulation the random variable X by a vector X = (X_1, …, X_n) of random variables X_1, …, X_n.

Replacing Y by $1_{[Y \in A]}$, for each event A, this definition of conditional expectation can be used to define a distribution of Y conditional on X = x, denoted by p_Y(|X = x) such that a conditional expectation of Y given X can be obtained computing, for each x, the expectation of Y with respect to this distribution of Y conditional on X = x.

In the discrete case, the distribution of Y conditional on X = x is given by

$$p_Y[A|X = x] = p(X = x,\ Y \in A)/p(X = x).$$

For continuous random vectors, the conditional density of Y given X is, for each real x, the density of the conditional distribution of Y given X = x, that is the real function $f_{Y|X=x}$ such that, for $F_{Y|X=x}$ denoting the cumulative distribution function of the conditional distribution of Y given the event [X = x],

$$F_{Y|X=x}(y) = \int_{-\infty}^{y} f_{Y|X=x}(u)du.$$

A better understanding of the information given by the conditional expectation on the correlation between different variables is obtained by computing the variance of the conditional expectation as the variance of a random variable whose variation is limited by eliminating all dispersion of Y within any set determined by a fixed value of X. The variance of Y can be decomposed into the sum of this variance of the conditional expectation and the expectation of another variable, the variance of the distribution of Y conditional on X.

$$VY = V(E(Y|X) + E(V(Y|X),$$

for V(Y|X) denoting the function that associates to each x the variance of the distribution with cdf $F_{Y|X=x}$.

A.7 Basic Distributions

This section brings examples of probabilities of each kind that will be useful in the development of the probabilistic determination of preferences.

A.7.1 Examples of Discrete Probability Distributions

A.7.1.1 The Bernoulli Distribution

The simplest sample space is that of the occurrence or not of a well specified fact, like success in an experiment or acceptance of a proposal. The space of events has then only 4 elements: {{yes}, {no}, Φ, S = {yes, no}}. A random variable has a Bernoulli distribution when its range has only two values 1 and 0, 1 associated to the occurrence of a given event of probability q and 0 associated to its complement. Thus

$$p_X(x) = q \text{ for } x = 1,\ p_X(x) = 1 - q \text{ for } x = 0 \text{ and } p_X(x) = 0 \text{ for any other real } x.$$

This implies that the cdf of X assumes the value 0 in R^-, 1−q in the interval [0,1) and 1 otherwise.

The expected value of a random variable with the Bernoulli distribution 1_A is p(A). Since $1_A^2 = 1_A$, the variance is

$$V(1_A) = p(A)(1 - p(A)).$$

Thus, for small p(A), $V(1_A)$ is slightly smaller than $E(1_A)$.

A.7.1.2 Binomial Distribution

The binomial distribution is the distribution of the sum of a number of Bernoulli variables with the same probability of the value 1. Two parameters determine a binomial distribution: the probability q of 1 in each of the Bernoulli summands and the number n of summands.

The classical example of a random variable with binomial distribution is the number of successes in a number of independent trials each of them with the same probability. Let S be the set of possible sequences of results of tossing n times a coin that has the probability q of showing the side with a head and let X be the number of heads observed. X has a binomial distribution with parameters n and q ($X \sim \text{Binomial}(n,q)$).

Since the expected value of a sum of random variables is the sum of the expected values of the summands,

$$EX = nq.$$

By independence,

$$V(X) = nq(1 - q).$$

Employing independence and a simple combinatorial computation, it may also be proved that, for every integer k from 0 to n,

$$p[X = k] = \{n!/[k!(n - k)!]\}q^k(1 - q)^{n-k}.$$

From this and Newton Binomial Formula for the power of a sum follows

$$\sum_{k=0}^{n} p[X = k] = 1.$$

A.7.1.3 Poisson Distribution

X has a Poisson distribution in the sample space N of the natural numbers, with parameter λ ($X \sim \text{Poisson}(\lambda)$) for a positive real $\lambda \leftrightarrow$

$$p[X = k] = e^{-\lambda}\lambda^k/k!, \text{ for every } k \in N.$$

An example of random variable with Poisson distribution is given by the number of particles emitted by a radioactive source in a given time.

Notice that $\sum_{k=0} \lambda^k/k! = e^{\lambda}$, so that, in fact, if X has a Poisson distribution, then

$$\sum_{k=0}^{n} p[X = k] = 1.$$

Notice also that $\lim_{n\to\infty}\left(1+\frac{\lambda}{n}\right)^n = e^{\lambda}$. This implies that,

for $X_n \sim \mathrm{Binomial}(n, \lambda/n)$ and $X \sim \mathrm{Poisson}(\lambda)$, $\lim_{n\to\infty} p[X_n = k] = p[X = k]$.

For $X \sim \mathrm{Poisson}(\lambda)$,

$$E(X) = \lambda$$

and

$$V(X) = \lambda.$$

A.7.2 Examples of Continuous Probability Distributions

A.7.2.1 Uniform Distribution

X has a Uniform distribution on the interval [a, b] ($X \sim \mathrm{Uniform(a,b)}$) $\leftrightarrow$ X has the density f_X determined by

$$f_X(x) = [1/(b-a)]1_{[a,b]}(x) \quad \text{for } a < x < b \text{ and 0 otherwise.}$$

That means, the density of x depends only on x being inside or outside the interval [a, b]. So, this density has a rectangular graph and

$$EX = (a+b)/2.$$

A.7.2.2 Triangular Distribution

X has a triangular distribution on the interval [a, b] with mode M, for $M \in [a, b]$ ($X \sim \mathrm{Triangular(a,M,b)}$) $\leftrightarrow$

$$f_X(x) = [(x-a)/(M-a)] * 2/(b-a) \text{ for } a < x \leq M,$$
$$f_X(x) = [(b-x)/(b-M)] * 2/(b-a) \text{ for } M \leq x < b$$

and

$$f_X(x) = 0 \text{ otherwise.}$$

If $X \sim \mathrm{Triangular(a,M,b)}$, the mode of X is M and if M = (a+b)/2, then X is symmetric around M, that is with expected value, mode and median equal to M. The expectation of X is

$$EX = (a+b+M)/3.$$

A.7.2.3 Exponential Distribution

$$X \sim \text{Exponential}(\alpha) \leftrightarrow f_X(x) = \alpha e^{-\alpha x} 1_{R^+}(x).$$

For X with an exponential distribution,

$$EX = \alpha^{-1}.$$

A.7.2.4 Pareto Distribution

$$X \sim \text{Pareto } (\alpha, \beta) \leftrightarrow f_X(x) = [\beta\alpha^\beta/(x^{\beta+1})]1_{[\alpha,+\infty)}(x) \text{for } \alpha > 0 \text{ e} \beta > 0.$$

For X with a Pareto distribution,

$$E(X) = \alpha\beta/(\beta - 1), \text{ for } \beta > 1,$$

and

$$E(X) = +\infty, \text{for } \beta \leq 1.$$

$$X \sim \text{Pareto } (\alpha, \beta,) \leftrightarrow \ln(X/\alpha) \sim \text{Exponential}(\beta).$$

A.7.2.5 Normal Distribution

X has a normal distribution with location parameter μ and dispersion parameter σ $(X \sim \text{Normal}(\mu, \sigma^2)) \leftrightarrow f_X(x) = (\sqrt{2\pi})^{-1}\sigma^{-1}\exp[-(1/2)(x - \mu)^2/\sigma^2]$.

$$EX = \mu.$$

$$\text{Var}(X) = \sigma^2.$$

The normal distribution has the following useful properties:

$$X \sim \text{Normal}(\mu, \sigma^2) \leftrightarrow (X - \mu)/\sigma \sim \text{Normal}(0, 1);$$

any linear combination of random variables with a normal distribution has a normal distribution;

if X_n is the sum of n random variables independent and identically distributed with a distribution with expected value μ and standard deviation σ then the distribution of $(X_n - n\mu)/(\sigma\sqrt{n})$ approaches a Normal(0,1) distribution as $n \to \infty$.

As a particular instance of this last property, for the summands with Bernoulli distributions, we have

$$X \sim \text{Binomial}(n, q) \to (X - nq)/\sqrt{[nq(1 - q)]} \cong \text{Normal}(0, 1)$$

A.7.2.6 The Lognormal Distribution

X has a lognormal distribution $\leftrightarrow$ lnX has a normal distribution.
The density of the distribution of X is then

$$f_X(x) = (x\sqrt{2\pi})^{-1}\zeta^{-1}\exp[-(1/2)(\ln(x/\xi))^2/\zeta^2]1R_+(x),$$

for ξ the median of the distribution of X and ζ = Var(lnX).

$$EX = \xi \exp[(1/2)\zeta^2],$$
$$Var(X) = (EX)^2[\exp(\zeta^2) - 1].$$

A.7.3 Examples of Multidimensional Distributions

For joint distributions, specially useful is to consider joint normal distributions.

A.7.3.1 Bidimensional Normal Distribution

The pair of random variables (X, Y) has a bidimensional normal distribution with expected values μ_X and μ_Y, variances σ_X^2 and σ_Y^2 and correlation coefficient ρ_{XY}, what is formally denoted by (X,Y) $\sim$Normal($\mu_X, \mu_Y, \sigma_X^2, \sigma_Y^2, \rho_{XY}$) if and only if their joint density has the form

$$f_{XY}(x,y) = C_1\exp\{C_2[[(x-\mu_X)/\sigma_X]^2 - 2\rho[(x-\mu_X)/\sigma_X][(y-\mu_Y)/\sigma_Y] + [(y-\mu_Y)/\sigma_{Y]}]^2\}$$

where

$$C_1 = (2\pi\ \sigma_X\sigma_Y)^{-1}(1-\rho_{XY}^2)^{-1/2}$$

and

$$C_2 = [-2(1-\rho_{XY}^2]^{-1}.$$

So,

$$\text{if } \rho = 0,\ \text{X and Y are independent.}$$

This extends to vectors of random variables.

A.7.3.2 Multidimensional Normal Distribution

The vector of random variables $X = (X_1, \ldots, X_n)$ has a multidimensional normal distribution with expected values $\mu_1, \ldots, \mu_n$, variances $\sigma_1^2, \ldots \sigma_n^2$ and correlation coefficients ρ_{ij} for $i \neq j$ from 1 to n ($X \sim \text{Normal}(\mu,\Sigma)$) if and only if their joint density has the form

$$f_X(x) = [(2\pi)^n \det(\Sigma)]^{-(1/2)} \exp\{(-1/2)(x-\mu)^T \Sigma^{-1}(x-\mu)\},$$

where μ is the vector of coordinates μ_i and Σ is the matrix whose entry Σ_{ij} for $i \neq j$ is the covariance $\rho_{ij}\sigma_i\sigma_j$ and for $i = j$ is the variance σ_n^2.

A.7.3.3 Properties of the Multidimensional Normal Distribution

$$X \sim \text{Normal}(\mu,\Sigma) \leftrightarrow \text{for any matrix A, } (AX) \sim \text{Normal}(A\mu, A^T \Sigma A).$$

So,

$$X \sim \text{Normal}(\mu, \Sigma) \text{ for } \Sigma = A^T A, \text{ for a square invertible matrix } A \leftrightarrow$$
$$A^{-1}(X-\mu) \sim \text{Normal}(0, I), \text{ I denoting the identity matrix,}$$

and, in particular,

$$X \sim \text{Normal}(\mu, \Sigma) \text{ for } \Sigma = A^T A, \text{ for a square invertible matrix } A, \rightarrow$$

$A^{-1}X$ is a vector of independent coordinates.

The Chi-Squared distribution is the distribution of the sum of squares of normal distributions.

$$X \sim \aleph_n^2 \leftrightarrow X = \sum_n Z_i^2 \text{ for } Z_1, \ldots, Z_n \text{ i. i. d. Normal}(0, 1).$$

If $X \sim \aleph_n^2$, then $EX = n$,
so that,

$$\text{if } X_1, \ldots, X_n \text{ are i. i. d. Normal}(0, \sigma^2) \text{ and } X = \sum_n X_i^2, \text{ then}$$

$$E(X/\sigma^2) = n$$

and

$$E(X/n) = \sigma^2.$$

It is easy to generalize this definition to:

$$Y \sim \aleph_n^2 \leftrightarrow Y = (X-\mu)^T \Sigma^{-1}(X-\mu)$$

for $X \sim \text{Normal}(\mu, \Sigma)$, for a square invertible matrix Σ.

Independence of quadratic forms is the subject of Cochran Theorem, employed to determine the F distribution.

Cochran Theorem

For $X \sim \text{Normal}(\mu, I_{nXn})$, let $Y_i = (X - \mu)^T \Sigma_i (X - \mu)$ for Σ_i symmetric idempotent matrices of ranks r_i such that $\sum_{i-1}^{n} r_i = n$ and $\sum_{i-1}^{n} \sum_i = I_{nXn}$. Then the Yi are independent $\aleph^2 r_i$ random variables.

The distribution F is the distribution of a quotient:

$$F \sim F_{m,n} \leftrightarrow F = (X_1/m)/(X_2/n)$$

for independent $X_1 \sim \aleph_m^2$ and $X_2 \sim \aleph_m^2$.

A.8 Regression

The process of approximating the distribution of a random variable by a conditional distribution is known as regression.

For normal distributions, the conditional expectation of Y conditional on X is a linear function of X.

In the case of one-dimensional X, this linear regression function is given by

$$E(Y|X = x) = E(Y) + \sigma_X^{-1} \sigma_Y \rho_{XY} (x - E(X)),$$

so that the variance of the conditional expectation is $\rho_{XY} \sigma_Y^2$.

Also, in the normal case, the conditional variance does not depend on X and is given, for one-dimensional X, by $(1 - \rho_{XY}^2)\sigma_Y^2$.

Starting from a sample $((Y_1, X_1), \ldots, (Y_n, X_n))$ of n independent random vectors, the coefficients of this linear function can be approximated by those of the linear projection operator in the sense that, for $X = (X_1, \ldots, X_n)^T$ and $Y = (Y_1, \ldots, Y_n)^T$,

$$\hat{Y} = X(X^T X)^{-1} X^T Y \sim \text{Normal}\left(E(Y|X), X(X^T X)^{-1} X^T \sigma_Y^2\right).$$

From Cochran's Theorem follows that the two random variables $\left(Y-\widehat{Y}\right)^T\left(Y-\widehat{Y}\right)$ and $\left(\widehat{Y}-\overline{Y}\right)^T\left(\widehat{Y}-\overline{Y}\right)$ are independent with distributions $\aleph^2_{n-m}$ and $\aleph^2_{m-1}$, respectively, for m the number of coordinates of each X_i, so that

$$\left(\left(\widehat{Y}-\overline{Y}\right)^T\left(\widehat{Y}-\overline{Y}\right)\Big/(m-1)\right)\Big/\left(\left(Y-\widehat{Y}\right)^T\left(Y-\widehat{Y}\right)\right)\Big/(n-m)\sim F_{m-1,n-m.}$$

Index

A. Parracho Sant'Anna, *Probabilistic Composition of Preferences, Theory and Applications*, Decision Engineering,
DOI 10.1007/978-3-319-11277-0

MIX
Papier aus verantwortungsvollen Quellen
Paper from responsible sources
FSC® C105338
FSC
www.fsc.org

If you have any concerns about our products,
you can contact us on
ProductSafety@springernature.com

In case Publisher is established outside the EU,
the EU authorized representative is:
Springer Nature Customer Service Center GmbH
Europaplatz 3, 69115 Heidelberg, Germany

Printed by Libri Plureos GmbH
in Hamburg, Germany